Plastic
Deformation
of
Nanostructured
Materials

Plastic Deformation of Nanostructured Materials

A. M. Glezer
E. V. Kozlov
N. A. Koneva
N. A. Popova
I. A. Kurzina

CISP

CRC Press
Taylor & Francis Group
Boca Raton London New York

CRC Press is an imprint of the
Taylor & Francis Group, an **informa** business

CRC Press
Taylor & Francis Group
6000 Broken Sound Parkway NW, Suite 300
Boca Raton, FL 33487-2742

First issued in paperback 2020

© 2017 by CISP
CRC Press is an imprint of Taylor & Francis Group, an Informa business

No claim to original U.S. Government works

ISBN-13: 978-0-367-57320-1 (pbk)
ISBN-13: 978-1-138-07789-8 (hbk)

Visit the Taylor & Francis Web site at
http://www.taylorandfrancis.com

and the CRC Press Web site at
http://www.crcpress.com

Contents

Introduction

The mechanical properties of materials are the basis of technical progress in various areas of engineering, instrument making, aerospace technology, nuclear technologies and many other aspects of human activity. Recently, the effort of materials scientists and other experts has resulted in the development of new approaches to increasing the strength and ductility of advanced structural materials. A special position is occupied by nanotechnologies capable of changing qualitative the mechanical behaviour of nanomaterials as a result of extensive dispersion of the structure. It has been shown that a large decrease of the grain size of the polycrystalline materials and the corresponding large increase of the bulk density of the grain boundaries greatly change the behaviour of dislocations – the main carriers of plastic flow and, consequently, the mechanisms of elastic deformation and the associated mechanical properties. A large number of scientific articles concerned with the effect of the small size of the crystals on the dislocation mechanisms of plastic deformation has been published in recent years. Undoubtedly, it is now time to publish the results and generalise the most important data.

The Russian literature contains a number of excellent monographs written by Russian and foreign scientists and devoted to the dislocation and disclination physics of plastic deformation of polycrystalline materials. Without pretending that this list is complete, one should mention first of all the monograph by R.W.K. Honeycombe 'Plastic deformation of materials' (1972), V.I. Trefilov, Yu.V. Mil'man and S.A. Firstov 'Physical fundamentals of the strength of refractory metals' (1975), V.E. Panin 'Structural levels of deformation of solids' (1985), V.V. Rybin 'High plastic strains and fracture or metals' (1986), O.A. Kaibyshev and R.Z. Valiev 'The grain boundaries and properties of metals' (1987), M.A. Shtremel' 'The strength of alloys' (in two parts, 1997) and a number of others.

Unfortunately, these monographs do not pay attention to the evolution of the mechanisms of dislocation flow at relatively small grain sizes (smaller than 1 μm). In the list of the monographs and reviews it is important to mention in particular books by K. Koch 'Constructional nanocrystalline materials' (2012) and M.Yu. Gitkin and I.A. Ovid'ko 'Physical mechanics of deformed nanostructures' (2003), and also a review by R.A. Andrievskii and A.M. Glezer 'The strength of nanostructures' in the journal Uspekhi fizicheskikh nauk (2009). The first of them is concerned mainly with the methods for producing constructional materials, their thermal stability and description of the mechanical and corrosion properties. As regards the plastic deformation mechanisms of nanomaterials, they are studied mostly from the viewpoint of computer simulation and do not describe the relationships observed in actual experiments. The latter also applies to a large extent to the second of the previously mentioned publications.

What is the subject of this book? The book examines in detail and systematically the special features of the mechanical behaviour and corresponding structural mechanisms of the behaviour of crystal structure defects with a decrease of the grain size in a polycrystalline ensemble in a stage preceding the nanolevel (from 1 μm to 100 nm) and in the nanosize range (less then 100 nm). Attention is given to the deformation behaviour of 'large' nanocrystals using the terminology proposed in [1] when the plastic deformation takes place by the nucleation, interaction and annihilation of the dislocations, up to 'middle sized' nanocrystals where the controlling role is played by the processes of grain boundary sliding.

This book is the result of 20 years of joint studies by researchers in Tomsk and Moscow, concerned with the strength of materials. The first chapter examines the stages of strain hardening of the polycrystals having different crystal lattices, and the effect of the dimensional factor on this process. The strain hardening pattern is examined on two structural levels (microscopic and the so-called mesoscopic level). The chapter also describes the condition of transition from dislocation sliding to twinning and martensitic transformation. The second chapter generalises the relationships governing the formation of dislocation structures in the deformed polycrystals and describes the mechanical properties under the effect of the change of the grain size (the Hall – Petch relation and its anomaly). Special attention is given to the conditions of transition from dislocation slip to grain boundary sliding with a decrease of the grain size. The third chapter contains the results of a detailed analysis of the main components of the dislocation structure from the viewpoint of the geometrically necessary and statistically stored dislocations. The concept of the critical grain size is introduced. The role of the inclusions of the second phase is evaluated. The fourth chapter examines the internal stress fields formed during the dislocation plastic flow. The methods for evaluating these fields and special features of the evolution in dependence on the grain size are outlined. The fifth chapter is concerned with the nature of high strain (severe) processes actively studied at present. The 'roadmap' of possible structure formation processes, observed at gigantic plastic strains, is described for the first time in the scientific literature. The important role of the cyclic processes of low-temperature dynamic recrystallisation and phase transformations, including amorphisation and crystallisation during deformation, is stressed. The sixth chapter uses titanium as an example to describe the structure and mechanical properties of modified surface layers of materials produced by ion implantation. It is shown that the target can be greatly strengthened by developing nanostructured phases of different nature formed during implantation with different ions.

We believe that this monograph can fill the existing gap in the publications concerned with the structural mechanisms of plastic deformation of ultrafine-grained and nanostructured materials which are of considerable scientific and applied interest at the moment. We hope that the book will be useful to scientists, engineers, post graduate students and others working in the problems of physics of strength and development of highly efficient constructional multifunctional materials.

We would be happy to receive any comments and wishes directed at improving the quality of the book and its importance for advanced materials science.

Reference

1.   Glezer A.M., Structural classification of nanomaterials. Deformatsiya i razrushenie materialov. 2010. No. 2. 1-18.

Stages of plastic deformation of polycrystalline materials

1.1. Introduction. Description of the problem

Different plastic deformation processes (tensile loading, compression, rolling, extrusion, creep, fracture) are usually characterised by distinctive stages. Of these stages of active deformation, uniaxial tension and compression have been studied most extensively. These types of deformation have been investigated widely on single and polycrystals with different grain sizes and the type of crystal lattice. Many dependences of the stress (σ) on strain (ε), $\sigma = f(\varepsilon)$, in the true coordinates have been published. A system of views of the individual stages of deformation under tensile and compressive loading has been formed. Although the relationships of work hardening have been studied for a long time, these problems still remain in the centre of attention of the world society of metal physicists. This is indicated by the fact that in the 11th volume of Dislocation in Solids, published in 2002 (edited by F.R.N. Nabarro M.S. Duesbery), a large part of the reviews was concerned with work hardening. In 2003, the series Progress in Materials Science included a review by U.F. Kocks and H. Mecking of this problem.

In this chapter, we analyse the current views regarding the individual stages of plastic deformation and work hardening of the polycrystals. The stages of deformation of the polycrystalline metallic materials were generalised for the first time in the well-known monographs [1–4]. The eight-stage pattern of plastic flow of the metallic polycrystals has been experimentally determined and described. The stages differ both in the value of the work hardening coefficient

$$\theta = \frac{d\sigma}{d\varepsilon}, \tag{1.1}$$

and in the dependence of this coefficient on ε [5]. The studies carried out to determine the individual stages of the polycrystalline materials include the work of a large group of foreign and Russian investigators. Unfortunately, the materials presented in the currently available Russian textbooks for plastic deformation of the polycrystals is still very scarce and no attention is given to the actual problem of the plastic deformation stages. The textbooks and lecture course literature still describe the old-fashioned three-stage deformation pattern and, in most cases, only for single crystals. The stages of work hardening of the polycrystals have practically not been investigated. At the same time, these problems are in the centre of attention of the world society of experts in metal physics and strength of materials.

1.2. Main stages of plastic deformation of polycrystals at the mesolevel

In the generalised form, the characteristics of the individual stages of deformation of polycrystalline aggregates in tensile or compressive loading are shown in Figure 1.1. In particular, the Figure shows the

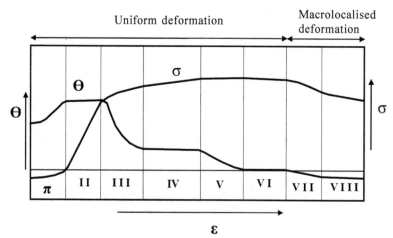

Fig. 1.1. Plastic deformation stages. The dependences $\sigma = f(\varepsilon)$ and $\theta = f(\varepsilon)$ are shown. The Roman numbers indicate the plastic deformation stages. The ranges of uniform and macrolocalised deformation are shown.

ranges of uniform deformation and macrolocalised deformation. The latter starts when the Backofen–Konsider condition is satisfied [6]:

$$\sigma = \theta. \tag{1.2}$$

This effect may occur in tensile deformation and, consequently, a 'neck' forms in the specimen [7]. In compression deformation, the fulfilment of the Backofen–Konsider condition may lead to macro-localisation, formed in strictly localised 'shear' through the entire specimen [7].

In a general case, at high plasticity one may observe eight stages of plastic deformation (Fig. 1.1) up to the start of fracture. In a number of cases the individual stages are not observed. The even and odd stages in Fig. 1.1 for the mechanical behaviour differ from each other. In the even stages, the work hardening coefficient θ is constant. In other words, the dependence $\sigma = f(\varepsilon)$ is linear in each even stage. In the odd stages, the value of θ decreases, and the dependence $\sigma = f(\varepsilon)$ in these cases is non-linear.

The value of θ is the highest in stage II and in decreases when each subsequent even stage stars. In the stages II and IV $\theta_{II} > 0$ and $\theta_{IV} > 0$, respectively, in the stage VI $\theta_{VI} = 0$, and in the stage V_{III} $\theta_{VIII} < 0$. The following sequence may be written:

$$\theta_{II} > \theta_{IV} > \theta_{VI} > \theta_{VIII}. \tag{1.3}$$

The relationship (1.3) is efficiently satisfied in deformation of different materials when the appropriate stages are reached. The change of the grain size may exclude some stages, but the relationship (1.3) remains unchanged.

1.3. Determination of the plastic deformation stages in FCC metals and solid solutions

The individual stages of the flow curves were detected for the first time in studies by G. Sachs and I. Weerts [8]. In the following seven decades, the problem of the individual stages was intensively studied (early history of the problem was described by the authors of this book in [9]). Initially, A. Zeeger [10] used single crystals to describe the well-known three-stage scheme of work hardening. However, already in 1990 the four-stage pattern of hardening for single and polycrystals of FCC metals during multiple slip in them was reliably established [9]. This dependence $\sigma = f(\varepsilon)$ is typical of single crystals close to the [001] orientation, and polycrystals in a wide grain size range. Figure 1.1 shows the transitional stage (indicated by π), the stage II of linear hardening with a high value of $\theta = d\sigma/d\varepsilon$, the stage III of parabolic hardening and the long stage IV with linear hardening, and also the stage II but with a low value of θ. The subsequent even stages are stages with linear hardening, and in the odd stages the work hardening coefficient decreases with strain. Recently, special attention has been paid to the behaviour of the FCC materials in the stages V and VI. Here there are several points of view. The most realistic view is based on the fact that in these stages the deformation stress reaches saturation (see the review in [11]]. The work hardening coefficients θ_{VII} and θ_{VIII} in the stages of macrolocalised deformation have negative values.

As already mentioned, the Russian textbooks for the mechanical properties of the materials still describe the three-stage dependence $\sigma = f(\varepsilon)$. At the same time, already in 1998, at the Eighth International Conference on the Strength of Metals and Alloys (ICSMA-8, Tampere, Finland) a special session was devoted to physical processes taking place in stage IV (see, for example, [12]). The stage IV is the longest stage in the $\sigma = f(\varepsilon)$ dependence (Fig. 1.1). This stage was described in the middle of the 20th century without identification. A large number of examples can be found in [1,2,4,9]. The main processes of mechanical treatment of the materials take place in particular in stage IV and, therefore, the investigation of the mechanisms of deformation and evolution of the structure in this stage is very important. Table 1.1 gives the list of authors of studies who investigated and described stages II, III and IV in the metallic polycrystals.

1.4. Some historical data for the determination of the stages II–IV of plastic deformation in polycrystalline materials

This chapter includes the critical analysis and generalisations of the patterns of work hardening of both meso- and sub-micropolycrystals. The current views on the individual stages of work hardening are reviewed. Attention is given to the role of structural transformations in the formation of the individual stages of plastic deformation. The changes in the deformation mechanisms, leading to the occurrence of new hardening stages, are described. At the same time, attention is given to the special features of the initial structure of the materials, leading to changes in the characteristics of the individual changes. In particular, special attention is given to the role of the grain size and the relationship of the grain size with the deformation mechanisms at the microlevel. Analysis was carried out using both the literature data and the results obtained by the authors of this book.

At present, special attention is given to the relationships governing the plastic deformation of metallic polycrystals with very fine grains. If previously the grain size of several microns was regarded as very small [13], now the scientists are producing metallic polycrystals with the average grain size in the range 1 μm–25 nm [14–16]. For some of these grains the stress σ–strain ε relationships were recorded and the stages of plastic flow, differing in the value of θ, were determined [17, 18]. Current studies describe the storage of information on the dependence $\sigma = f(\varepsilon)$ for ultrafine-grained polycrystalline materials. If the previously examined polycrystals were classified as being at the mesolevel, the currently studied polycrystals are classified as being at the microlevel. The grain size of 1 μm separates the mesolevel from the microlevel [19] (Table 1.2).

The first results in the measurement of the form of the $\sigma = f(\varepsilon)$ dependence for the polycrystals at the microlevel were partially obtained on the polycrystals of copper and aluminium and aluminium alloys. The largest number of the experiments were carried out using the copper polycrystals with different grain sizes.

1.5. Individual stages of plastic deformation in the BCC metals and alloys

This problem was extensively investigated in the 90s of the previous century. In a general case, the flow curves of the polycrystals of Fe and different steels include the hardening stages III and IV. This problem was investigated for martensitic and bainitic steels in Russian studies [20–22] and in studies of the German school of W. Dahl [23–27) for ferritic, ferritic–pearlitic and pearlitic steels. In the analysis of the individual stages, the authors of the above studies used, as in the case of the FCC metals, the $\theta = f(\varepsilon)$ or $\theta = f(\sigma)$ dependence. In this approach the stages III and IV can be efficiently separated. The authors of the Ukrainian school, who assumed that the $\sigma = f(\varepsilon)$ dependence is approximated by a set of parabolas [3, 28], could not identify these deformation stages.

Table 1.1.

Authors	Stages	Publication date
Bell J.F., Carreker R.P., Hibbard	I I I	1958-67
W.R. Hosford W.F.	I I I	60's
Schwink Ch., Macherauch E.	I I and I I I	1964
Marcinkowski M.J., Chessin H.N.		1964
Popov L.E., Kozlov E.V.,	I I and I I I	1965
Aleksandrov N.A.	I I and I I I	
Korotaev A.D., Koneva N.A.	I I and I I I	(70's-80's)
Tsypin et al.	I I , I I I and	1976
Ivanova V.S., Ermishkin V.A.	IV	1965
Van. Hein-Peter Stiwe	IV	1969
Landford G, Cohen M.	IV	1978
Lioyd D.J. et al.	IV	1981
Sevillano J. Gil et al.	IV	1985
Koneva N.A., et al.,	IV	1985
Brion M.G., Haasen P.	IV	1988
Zehetbauer M. et al.	IV	1988
Hughes D.A., Nix W.D	IV	1989
Trefilov V.I., et al.,	IV	1989
Kuhlmann-Wilsdorf D., Hansen N.	IV	1993
Bassim M.N., Liu C.D.	IV	1993
Fang X.F., Dahl W.	IV	1998
Panin V.E., et al.	IV	

Table 1.2. Classification of polycrystals according to the grain size

Scale level	Polycrystal type	$<d>$, average grain size
Mesolevel	Coarse-grained (macro) polycrystal	0.1...10 mm
	Conventional (meso) polycrystal	10...100 μm
	Fine-grained polycrystal	1...10 μm

Scale level	Polycrystal type	$<d>$, average grain size
Microlevel	Ultrafine-grained (UF) polycrystal	0.2...1 μm
	Submicrocrystals	50...200 nm
	Nanocrystals	3...50 nm
	Imperfect crystals and amorphous state	No grains

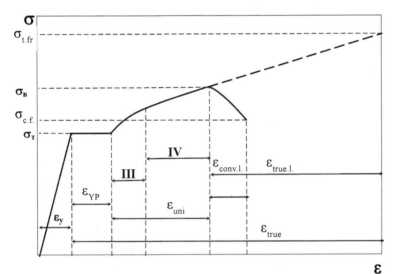

Fig. 1.2. Schematic representation of the $\sigma = f(\varepsilon)$ dependence for a ferritic steel polycrystal in uniaxial tension. Notations are explained in the text.

martensite and (or) bainite, the stage II is usually not observed.

However, this stage is found in the ferritic steels. This $\sigma = f(\varepsilon)$ dependence is shown schematically in the uniaxial tensile loading of a ferritic steel in Fig. 1.2. In addition to the stages III and IV, there is also the yield plateau ε_{YP} (the first region of localisation of deformation). It is followed by the range of uniform deformation $\varepsilon_{uni} = \varepsilon_{III} + \varepsilon_{IV}$. After reaching σ_B (the conventional fracture stress) there is the second stage (the neck) of localised deformation which in Fig. 1.2 is indicated in two measurements, $\varepsilon_{conv.l.}$ and $\varepsilon_{true.l.}$, i.e., correspondingly the conventional and true localised strains. Figure 1.2 also shows the conventional ($\sigma_{c.f.}$) and true ($\sigma_{t.fr}$) fracture stresses. To determine the true values of σ and ε in the strain localisation stages, it is necessary to measure carefully the parameters of the neck, in particular, the local reduction in area and elongation as carried out, for example, in [22]. Figure 1.2 also shows the yield limit σ_y, the range of quasielastic strain ε_e, and the total plastic strains ε_{true}. Thus, Fig. 1.2 combines schematically the stages of uniform and localised deformation. The transition from the stage of uniform deformation to the localised stage starts at the moment of fulfilling the Backofen–Konsider conditions (2) [6].

1.6. Storage of dislocations, internal stress fields and evolution of the dislocation structure

The main classic mechanism of plastic deformation is the dislocation deformation. The dislocations are generated by different sources, multiply, interact, form barriers, annihilate in the volume, transfer to the free surface and travel to the grain boundaries, accumulation to high densities in the volume of the material and forming different substructures. Consequently, as a result of electron microscopic studies, carried out on different FCC metals and solid solutions, it was established that all the investigated types of dislocation substructures (DSS) can be divided into two large classes [9, 29]: 1) the group of non-misoriented substructures, 2) the group of misoriented substructures. In the former there may be discrete misorientations, but they do not exceed 0.5°. In the group of these non-misoriented dislocation structures we can define (Fig. 1.3): a – the random distribution of the dislocations (1); b – pile-ups (2); c – uniform network structure (3); d – dislocation tangles (4); e – non-misoriented cells (5); and f – the cellular–network substructure (7).

The misoriented substructures (discrete misorientations at the sub-boundaries exceed 0.5°) include (Figure 1.4): a – the cellular substructure with the misorientation (6); b – cellular–network dislocation structure with smooth misorientations (8); c – microband substructure (9); d – the substructure with multi-dimensional discrete and continuous misorientations (10); e – fragmented substructure (11). There is another group of substructures, associated with twinning and martensitic transformations. The initial substructure in this group of substructures – the substructure with split dislocations (12) – can be related to non-misoriented substructures. The incomplete twinning results in the formation of the substructure (13) with multilayer stacking faults (f). They belong to the group of the misoriented substructures. The misoriented substructures also include twinned (14) (single-, two- or multi-dimensional) and a substructure with

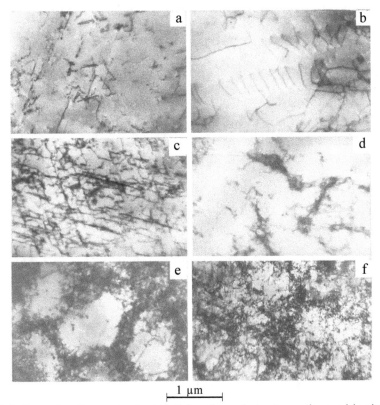

1 μm

Fig. 1.3. Example of non-misoriented dislocation substructures observed in single-phase FCC alloys: a) random distribution of dislocations; b) dislocation pile-ups; c) network dislocation substructures (DSS); d) tangles; e) cellular DSS; f) network–cellular DSS.

deformed martensite (15) (Fig. 2.2g and h, respectively). These substructures develop, fill the deformed material and transform to other substructures. The type of substructure, formed in the plastic deformation in the FCC metals and alloys, depends on the stacking fault energy (SFE), the test temperature (T), the magnitude of solid-solution hardening (τ_f) and the average scalar dislocation density ($<\rho>$).

Figure 1.5 shows schematically the sequence of the transformation of dislocation substructures and the corresponding stages of plastic deformation [9, 30–32]. This scheme describes the transformations in the dislocation structure of the FCC materials at moderate temperatures and active deformation. The pattern shown in the scheme relates exclusively to the single-phase materials and to the deformation carried out by total dislocations. The twinning and martensitic transformation are outside the range of this scheme.

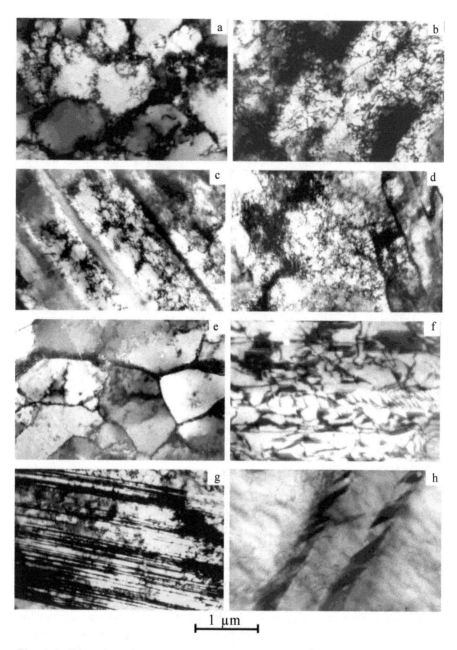

1 μm

Fig. 1.4. Examples of misoriented dislocation substructures, found in the FCC alloys of solid solutions: a) the misoriented cellular substructure; b) cellular–network substructure with smooth misorientations; c) microband substructure; b) the substructure with discrete and continuous misorientation; c) fragmented substructure; f) multilayer stacking faults; g) twin substructure; h) strain martensite.

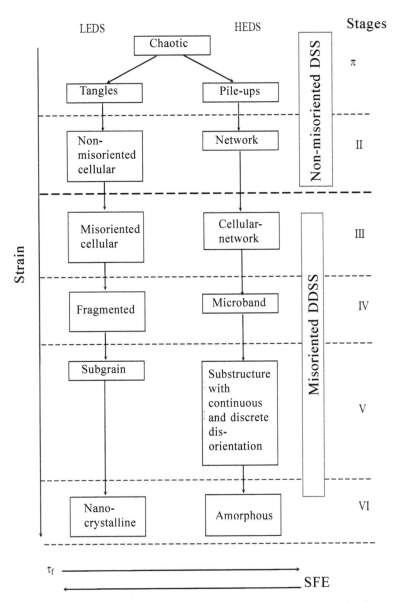

Fig. 1.5. Low-energy (LEDS) and high-energy (HEDS) sequences of substructural transformations, observed in the process of active deformation of the FCC materials at moderate temperature (schematic): DSS – the dislocation substructures, DDSS – dislocation–disclination substructures.

The existence of two sequences of substructural transformations has been established: low-energy (LEDS) and high-energy (HEDS)

[9, 30–32]. The information on the energy types of the dislocation substructures – LEDS, MEDS and HEDS (low-, medium- and high-energy structures) was presented in [33] (see also the review in [34] and discussed in detail at four international conferences dealing with the low-energy dislocation structures, organised in 1986, 1989, 1992 and 1995. At present, the generally accepted aspect of this concept are the low-energy dislocation structures (LEDS) and also high-energy dislocation structures (HEDS). The intermediate structure (MEDS – medium energy dislocation structure) is classified only in a small number of cases at present. The LEDS takes place in pure metals and diluted solid solutions with a high stacking fault energy. HEDS takes place in the concentrated solid solutions (high solid solution hardening) and in materials with a low stacking fault energy (difficult cross slip).

Figure 1.5 shows the classification of the substructures. At the same time, Fig. 1.5 shows the relationship of the dislocation substructure with the ensemble of the polycrystalline grains and their boundaries. Along the LEDS there is a sequence of a distinctively developed dislocation substructure by the formation of different types of boundaries. The cell boundaries transform to the boundaries of the fragments, then sub-grains, and finally the grain boundaries. Initially, the misorientation is produced by the excess dislocations of the same type piled up at the grain boundaries. Subsequently, partial dislocations form at the contact of the boundaries and then control the formation of the structure of the intergranular boundaries. In the LEDS the dislocation sequences constantly both annihilate or are 'displaced' to the boundaries. In the HEDS the sequences are characterised by a completely different process, in addition to the formation of the incomplete boundaries, a high density of scalar and excess dislocations is built up between them as a result of the high value of the solid solution hardening. Therefore, the HEDS sequence is completed by amorphisation. Of course, any of the sequences may, under specific conditions, be interrupted by the fracture process as a result of the formation of microcracks at the boundaries. The distribution of the dislocations in the volume of the material and the formed dislocation configurations determine the level of the internal stresses.

The dislocation ensemble contains several types of dislocation configurations. The most complete set of these configurations is shown in Fig. 1.6. The first set includes the individual dislocations and their pile-ups. The second set includes spatial configurations and

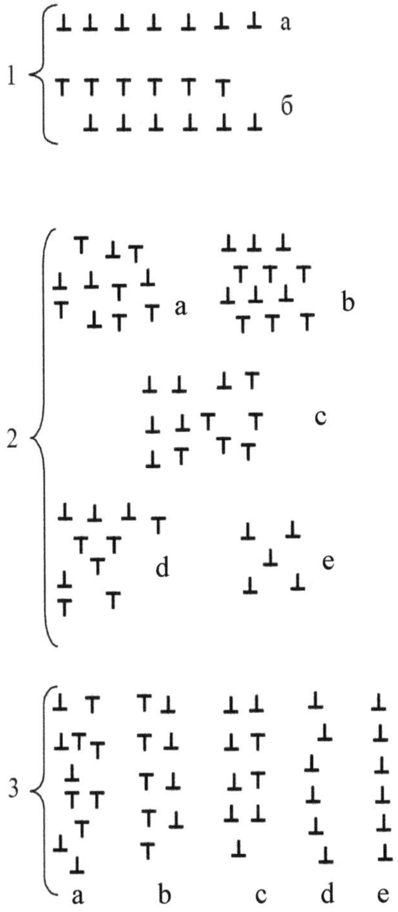

Fig. 1.6. Types of local dislocations configurations: 1) pile-ups (a – single, b – the multipolar); 2) dislocation groups (a – disordered, b – multipolar, c – polarised, d – partially charged, d – fully charged); 3) the walls of the cells (a – disordered, non-misoriented wall, b – ordered, non-misoriented wall, c – ordered misoriented wall, d and e – the boundaries of the fragments (d – imperfect boundary, e – perfect boundary)).

boundaries of the cells and sub-boundaries. In most cases, they are relatively incomplete and contain different numbers of dislocations of different size. If the boundary is not closed, then in the vicinity of the break in the boundary the amplitude of the field of the internal stresses is especially high. The broken sub-boundary is the nucleus of the partial disclination. In addition to this, the partial disclinations are localised in the contact of the imperfect boundaries of the cells and fragments.

The developed dislocation structure is characterised by the distribution of the defects in which the elastic fields of the defects greatly compensate each other. In these cases, we are usually concerned with the screening of the elastic fields of the defects and the screening radius R is introduced. In a comparatively uniform dislocation structure the screening radius is usually proportional to $\rho^{-1/2}$, and in the cellular structure its value is of the order of the cell size or the width of the walls of the cells [31]. In the presence of more powerful sources, such as partial disclinations, the screening radius may be considerably greater than for the purely dislocation sources, for example, of the order of the distance between the disclinations [35]. In the case of the complicated substructures this problem requires a separate special investigation [36].

The evaluation of the contributions of the initial stress fields to the deformation resistance is a complex task. The results of a small number of evaluations show that the contribution of the elastic fields of the dislocation ensemble to the deformation resistance is 1/3–1/5 of the acting stresses [36, 37]. This contribution should depend on the type of substructure formed in the appropriate stage of the plastic flow, the density and the distribution of the main types of defects, forming the internal stress fields. In particular, these defects include the dislocations and partial disclinations.

1.7. Evolution of the substructure – the basics of the physics of stages in gliding of total dislocations

The nature of the change of the stages in active plastic deformation has been discussed in a number of investigations. In these investigations, the authors stress different aspects of this phenomenon [38]. They include: 1) the change of the nature of gliding (from planar to spatial) and of the size of the shear zone; 2) the change of the number of acting gliding systems; 3) the transition from that simple storage of dislocations to the rearrangement of the dislocation structure and the increase of the fraction of the annihilating dislocation; 4) the inclusion of the rotating deformation modes assuming the development of continuous and discrete misorientations; 5) substructural transformations, i.e., the change of one substructure to another and the associated different mechanisms of inhibition of the dislocations. When analysing the $\sigma = f(\varepsilon)$ dependence, it is necessary to take all these effects into account. At the same time, the fifth approach, which actually includes all the previously mentioned approaches, appears to be most efficient. In the majority of studies, concerned with the nature of the individual stages, special attention is given to this direction [9, 39–41].

The transitional stage and stage II (Figs. 1.1 and 1.5) occur in the non-misoriented substructures. The stages III–IV are realised in the formation of the misoriented substructures. The LEDS sequence is characterised mainly by the discrete misorientations (deformation sub-boundaries [42–46]). An important role in the development of substructures is played by the evolution of misorientation with deformation. The HEDS sequence forms as a result of the co-existence of continuous and discrete misorientations [9]. In the stages IV and V in addition to the dislocation mechanisms of the formation of substructures an important role is starting to be played by the long-range stress fields and disclination mechanisms [47–49]. In particular, this involves the dipoles of partial disclinations and disclination loops [50]. It is important to stress that in theses stages of formation of the substructures it is quite easy to separate the elastic long-range and short-range effects, i.e. two different components as regards the screening radius.

The change of the substructures includes a change of stages. The beginning of each stage is closely linked with the formation of a new

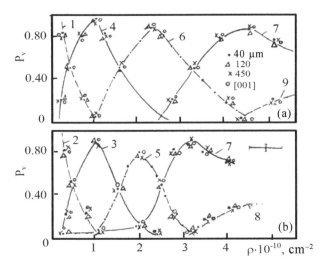

Fig. 1.7. Dependence of the volume fraction (P_V) of the substructures of different type on the scalar density of the dislocations in the poly- and single crystals of Ni_3Fe alloy with the long-range (a) and short-range (b) order. The figure shows the grain size of the polycrystals and the orientation of the single crystal. The numbers 1...9 indicate the types of substructure: 1 – the chaotic distribution of the dislocations, 2 – pile-ups, 3 – non-misoriented network substructure, 4 – non-misoriented cellular substructure, 5 – misoriented network substructure, 6 – misoriented cellular substructure, 7 – microband substructure, 8 – substructure with continuous and discrete misorientations, 9 – fragmented substructure.

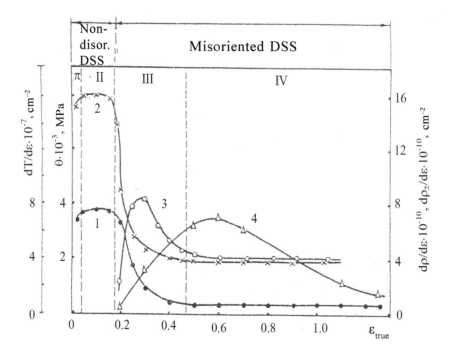

Fig. 1.8. Dependence of the work hardening coefficient θ (1), the rate of storage of the scalar $dp/d\varepsilon$ and excess $dp_\pm/d\varepsilon$ (3) dislocation density and the storage of ruptured sub-boundaries $dT/d\varepsilon$ (4) on the true strain (ε_{true}) of the Ni$_3$Fe ordered alloy. The dotted line indicates the stages of deformation.

type of dislocation substructure. This is clearly illustrated in Fig. 1.7 [9]. Here the volume fractions (P_V) of the appropriate substructures are presented in dependence on the average scalar density ($<\rho>$) of the dislocations. The decrease and increase of the volume fractions of the appropriate types of substructures characteristic of each stage are clearly visible. The end of each stage coincides with the disappearance of the previously formed substructure, the maximum of the volume fraction of the existing substructure and the formation of a new substructure. The scalar density of the dislocations is the controlling parameter of the substructural transformations [39,40]. This follows from Fig. 1.7 which shows the data for the dislocation density for three sizes of the grains and the single crystal [001] of the Ni$_3$Fe alloy. The values of P_V for them fit the same curve.

Quantitative estimates of the rate of storage of different types of defects at different stages are shown in Fig. 1.8 [50]. In stage II the rate of storage of the scalar dislocation density reaches the maximum value. The main mechanism of generation of the dislocations in stage II in the polycrystals is the generation at the grain boundaries (from microledges at these boundaries), and in the single crystals – generation from the free surface.

The transition to stage III marks the start of high-rate transverse gliding, the generation of dislocations in the volume of the material and the high-rate annihilation of dislocations. Consequently, in stage III the rate of storage of the excess dislocation density $d\rho_{\pm}/d\varepsilon$ is maximum [50,51[. It should be noted that $\rho_{\pm} = \rho_{+}-\rho_{-}$ Here ρ_{+} and ρ_{-} are the densities of dislocations of different sign since $\rho = \rho_{+}+\rho_{-}$. Stage III is the start of the storage of partial disclinations situated at the end of the ragged deformation sub-boundaries. The rate of this process is maximum in stage IV. The density of the sub-boundaries rapidly increases through the entire stage IV. Stage V and, in particular, subsequent stages, have not been studied sufficiently so far.

1.8. Transition to twinning and deformation martensitic transformation as an important factor of formation of stages of work hardening

In addition to the gliding of total lattice dislocations, the FCC materials may show twinning, sliding at the grain boundaries and the strain martensitic transformations of the $\gamma\to\alpha$ and $\gamma\to\varepsilon$ type. Examples of multistage curves in activation of new deformation mechanisms are shown in Fig. 1.9. In Fig. 1.9a the formation of the new stage II (transition from stage II_1 to stage II_2) in FCC polycrystals of 1.1C–13Mn–Fe steel is determined by the activation of the twinning mechanism [52]. Figure 1.9b shows that the Fe–Ni–Cr FCC steel is characterised by consecutive activation of several new deformation mechanisms from those described above. The letters (α, ε) denote the $\gamma\to\alpha$ and $\gamma\to\varepsilon$ martensitic transformations. Each such activation is accompanied by the formation of a new stage. Figure 1.9 shows the equally important role played in the formation of new deformation stages of substructure transformation and the activation of new deformation mechanisms.

1.9. Localisation of deformation – another reason for the formation of new stages

Any type of localisation of deformation leads to the formation of

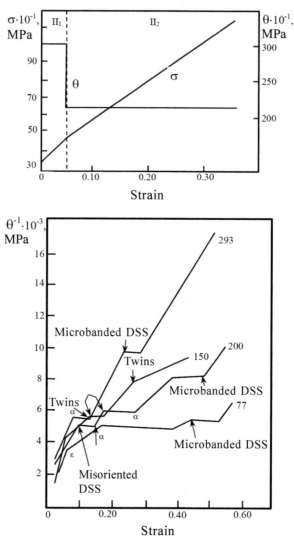

Fig. 1.9. Stages of deformation when the deformation mechanism of the FCC steels changes: a – 1.1C–13Mn–Fe steel. There are two stages II: II$_1$ and II$_2$ with jump-like variation of θ; b – Fe–Ni–Cr steel, deformation at different temperatures (the numbers at the curves). When the stages change, the new deformation mechanisms are indicated. In other cases, the formation of the new stages of hardening is associated with substructural transformations. The letters α and ε indicate γ–α and γ–ε strain martensitic transformations.

a new stage of plastic deformation in the σ–ε dependence. Several examples are shown in Fig. 1.2. Firstly, it is the yield plateau, secondly the sections on the flow curve, corresponding to the formation of the 'neck'. Sometimes, tensile loading of the specimen is characterised by the consecutive appearance of several different necks [20].

The formation of a neck is accompanied by a change of the conditions of calculating the acting stresses and local strains. At the same time, the stress tensor in the active deformation sections changes. Taking these factors into account changes the form of the σ–ε dependence (Fig. 1.2).

1.10. Factors complicating the characteristics of the deformation stages in meso-polycrystals

The individual stages of the $\sigma = f(\varepsilon)$ dependence were identified on the basis of mechanical tests. The reliable determination of the stages requires qualified experiments – efficient testing machine, efficiently prepared specimens, multiple reproduction of the experimental results. At the same time, it is important to take into account the factors changing the characteristics of the stages or complicating their identification.

The number of deformation stages, the duration of the stages and other parameters may change with:

1. The change of the test temperature;

2. The change of the orientation of single crystals;

3. In the transition from the single crystal to polycrystalline material;

4. The multicomponent nature of the texture of the polycrystalline aggregate;

5. The simultaneous action of several deformation mechanisms.

The beginning of a new stage, as mentioned above, is determined in accordance with [9] by the appearance of a new substructure and the maximum amount of the previous substructure in the volume of the material. This relationship may be complicated by the **polycrystalline state** of the material. Usually, the new substructure forms in the areas with a higher dislocation density. There may be cases in which the critical dislocation density [9, 39] is not yet obtained in the volume of the grain but it is obtained in the boundary zone. The new substructure starts to form here, and the transition from one station to another on the $\sigma = f(\varepsilon)$ dependence becomes blurred. At the given mean grain size the grain size distribution function is obtained. If we take the large and fine grains from this distribution, they are characterised by different dislocation density, and the new substructures will form at different strains. This factor distorts the 'sharpness' of transition from one stage to another.

This effect is even stronger in polycrystals in the presence of different grain sizes. If the volume of the materials contains groups of fine and large grains, the individual stages will be represented by the sum of stages of these groups of grains, i.e., the behaviour of the material will be similar to the composite behaviour. The same effect is obtained if the material contains distinctive two texture components [53]. The activity of slip in the grain depends on the Schmid factor of the acting slip systems [54]. The Schmid factors are determined by the grain orientations. In this situation, each group is characterised by its stages and the general pattern is blurred.

The factor complicating the individual stages is also the initial structure of the material, with substructural or dispersion hardening (martensite or bainite in steels, dispersed particles, carbides, pearlite, etc). In this case, the duration of stage III is shortened, stage II does not form at all. The main part of deformation takes place in the stages IV and V. If the deformation is accompanied by the dissolution or formation of secondary phases, the characteristics of the individual stages may also become blurred.

In transition to the ultrafine grain size, the 'sharpness' of the pattern of the individual stages is influenced by the main types of grain boundaries, the presence of the impurity particles in quadripole sections, at triple lines and the grain boundaries, the ratio of the strength of the intragranular sections and the boundary sections, the activation of hardening mechanisms (solid solution, dislocation, dispersion, polyphase), with the contribution comparable with that of the grain boundary hardening mechanisms.

1.11. Effect of the mesograin size on the individual stages of plastic deformation

The main factors changing the form of the individual stages $\sigma = f(\varepsilon)$ with the change of the grain size are: 1) the size of the shear zone; 2) the fraction of the region of the grains containing the contact stresses from the adjacent grains; 3) the relative fraction of the dislocations, emitted from the grain boundaries or generated in the volume of the grain; 4) changes of the texture and the Schmid factors of the acting slip systems; 5) the change of the structure of the grain boundaries, in particular, the change of the fraction of the boundaries with a low value of Σ (Σ is the inverse density of the coinciding sites).

Table 1.3. Effect of the mesograin size on the individual stages of the $\sigma = f(\varepsilon)$ dependences in active plastic deformation of the FCC metals and solid solutions

No.	Type of polycrystal	Average grain size d	Characteristic stages of plastic deformation
1	Coarse-grained polycrystal	0.1...10 mm	The $\sigma = f(\varepsilon)$ dependence is intermediate between single and polycrystal dependences. The role of grain boundaries is not important. The transition stage may be characterised by reduced value of θ
2	Conventional polycrystal	10...100 µm	Typical six-stage $\sigma = f(\varepsilon)$ dependence; stages II, III, IV and VI are distinctive

The review data for this problem are systematised in Table 1.3. With the refining of the grain size the possibilities of work hardening initially improve, and the flow curves contains the well-developed stages II–VI. With the decrease of the average grain size $<d>$ and its approach to the range 5–1 µm the sharper pattern of the individual stages of the $\sigma = f(\varepsilon)$ dependence is complicated by the increase of the number of the acting slip systems in the individual grains. This is caused by the fact that the contact stresses, formed between the adjacent grains, usually extend to a distance of several microns. Therefore, the polycrystals with the mean grain size $<d>$ from units to tens of microns are characterised mainly by the effect of the slip systems with the maximum Schmid factor. This pattern is observed at least in the central part of the grain body. Only in the areas of the grains in the vicinity of the grain boundaries the contact stresses, formed between the grains because of the incompatibility of their deformation, lead to the formation in the boundary region of the effect of the slip systems with a small or zero Schmid factor [55–57]. Consequently, the density of the geometrically necessary dislocations [58], carrying out slip in the boundary region, increases. In the fine-grained polycrystals in which the grain size is comparable with the size of the region of the effect of intergranular contact stresses, there are usually different slip systems in operation, determined by both the external applied stresses and contact stresses [54–57]. In this case, the grain size $d = 1$ μm is critical. With a further decrease of d the high initial hardening decreases the possibility of further work hardening, and the individual stages are suppressed. In the presence of very fine grains the deformation mechanisms change, and the flow curves contains mostly the stage without hardening – the stage VI.

The main characteristics of the individual stages in uniaxial tensile and compressive loading of the meso-polycrystals have been determined [39–41]. It has been shown that the patterns in both the single and polycrystals have a number of similar features. The nature of the transition stage π, of the stages II, III and IV is identified for the case of slip by the perfect dislocations on the basis of substructural transformations. The problems of the stages V and VI have been discussed and clarified. We shall stress a number of factors determined by a decrease of the grain size leading to a change in the number of stages. The main of them are: localisation of deformation, the change of the deformation mechanisms, i.e., the transition from the slip by conventional dislocations to twinning and strain martensite, and activity of the diffusion mechanisms.

An important achievement was the investigation of the substructure after high strains and determination of the reasons for the formation of a number of problems in the physics of work hardening of polycrystals on the mesolevel. They include the determination of the critical strains after which the statistically stored dislocations with the density ρ_S and the geometrically necessary dislocations with the density ρ_G and forming the sub-boundaries become the main factors of further work hardening. An important role is the cellular, fragmented or microband substructure with its own characteristic dimensions between the sub-boundaries. The barrier effect is the hardening factor in the substructure. Further, the development of deformation together with discrete misorientation in a number of materials may be accompanied by the formation of dislocations associated with the excess density. This determines the formation of a long-range component of the elastic fields, in addition to the short-range component. The role of the long-range component becomes more important as a result of the development of dislocation–disclination substructures. At high strains the LEDS sequence is completed by the formation of nanostructures, the HEDS – by amorphisation.

1.12. Changes of the structure of the polycrystalline aggregate and the pattern of the deformation stages with a decrease of the average grain size

In the fine-grained polycrystals in which the grain size is comparable with the size of the region of the effect of intergranular contact stresses, there are usually different slip systems determined by both the external applied and contact stresses. Therefore, with further refining of the grains and the transition to the microcrystals and then further to nanocrystals, the pattern of work hardening may change. The classification of the types of polycrystals on the basis of the grain size is presented in Table 1.2. With a decrease of the grain size the work hardening coefficient in stage IV decreases, and the stage slowly transforms to stage VI or even VIII. The stages IV, VI and VIII differ in the magnitude and sign of the work hardening coefficient θ. In the stage IV $\theta_{IV} > 0$, in stage VI $\theta_{VI} = 0$, and in the stage VIII $\theta_{VIII} < 0$. The pattern of the possible variants of the plastic deformation stages with a decrease of the average grain size is shown in Figure 1.10. Table 1.4 shows the evolution of the stages of plastic deformation with the decrease of the average grain size [18].

The change of the pattern of the individual stages of deformation with the decrease of the grain size is determined by the change of the structure and boundaries of the grains, the internal stresses, texture, the size distribution of the grains and the deformation mechanisms [18]. The change of the structure of the polycrystal and the deformation mechanisms in the entire possible range of the average grain size is shown in Table 1.5.

In transition to the ultrafine-grained and nanocrystals, the pattern of deformation localisation becomes more complicated. If at the grain size of $d > 10$ μm the fulfilment of the Backofen–Konsider condition $\theta = d\sigma/d\varepsilon = \sigma$[59] is accompanied by the formation of a neck, then in the nanorange the localisation of deformation takes place by strain bands and high-speed gliding at the grain boundaries and even their groups. The pattern of work hardening is characterised by the presence of long stages with an almost constant value θ (Fig. 1.10). As the value of θ decreases, the extent of localisation of deformation increases.

It may be concluded that in the presence of the fine grains the stages II and III are retained, and the state IV ($\theta > 0$) is frequently replaced by the stage VI ($\theta = 0$) or even by the stage VIII ($\theta < 0$).

Table 1.4. Effect of the size of the micrograins on the individual stages of the dependence $\sigma = f(\varepsilon)$ in active plastic deformation of the FCC metals and solid solutions

Type of polycrystal	Average grain size d	Chacteristic stages of plastic deformation
Microcrystals	0.1–10 μm	Suppressed stage IV, stages II, III, V and VI form
Submicrocrystals	10–100 nm	Stages II, III and VI form
Nanocrystals	3–10 nm	Stage III suppressed, mainly stage VI or VIII form
Imperfect crystals and amorphous state	No grains	Mainly stage VI or VIII form

Table 1.5. The structure and deformation mechanisms with the change of the grain size of polycrystals

Classification	Scale or grain size	Defective structure after preparation of material	Deformation mechanisms
Pile-ups, amorphous state	0.5–2 nm	Amorphous structure. Mean coordination number differs from crystalline number. Delamination in concentration	Displacement of free volume and similar mechanisms
Imperfect crystals	2–3 nm	Imperfect order approaching crystalline. Atomic spacings differ from crystalline. Distorted lattice. Delamination in concentration. Start of formation of the grain structure	Displacement of free volume. Gliding at grain boundaries and phase boundaries. Rearrangement of quasicrystalline lattice. Formation of dislocations.
Nanocrystals	3–10 nm	Almost perfect lattice. Lattice parameter constant. Crystal boundaries not sharp.	Gliding on imperfect boundaries, including with free volume. Grain rotation. Start of dislocatino sliding from boundary to boundary. High internal stress fields.

Submicrocrystals	10–100 nm	Defect-free crystals, separated by highly defective boundaries. Disclinations at grain junctions	Elastic moduli depend on the crystal size. GBs and their junctions are sources of internal stresses. Sliding of individual dislocations inside grains. High-speed gliding at the GBs at low temperatures
	50–100 nm	Internal stresses	Total dislocations
Microcrystals	0.1–1 μm	Crystals may contain dislocations or even substructure (cells, fragments). Difference GBs: imperfect defective, close to perfect of general type and perfect special	Internal stress fields from boundaries, their junctions and sub-boundaries. Dislocation intragranular sliding. Grain boundary gliding with increase of grain size and perfection of the boundaries gradually slows down or is displaced to higher temperatures
Fine-grained polycrystal	1–10 μm	Relatively perfect grains. Annealing twins. Substructure. Boundaries with a small number of defects, mostly of general type, less frequently special	Dislocation slip, twinning, strain martensite. In deformation internal stress fields from adjacent grains, boundaries and their junctions penetrate through the entire grain and ensure activity of systems with any Schmid factor. As in microcrystals, dislocations are emitted by sources at GBs.

1.13. The main factors determining the stages of deformation and the value of the work hardening coefficient in the microrange

The first factor is the increase of the dislocation density ρ in transition to the microcrystals and the subsequent decrease of the dislocation

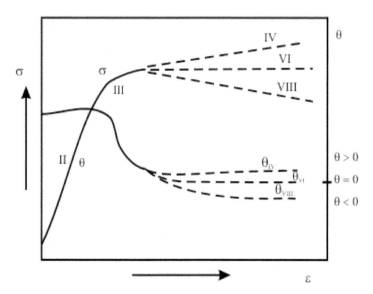

Fig. 1.10. Diagram of three possible charts of work hardening (softening) at high strains of the FCC polycrystals at the microlevel.

density in transition to sub-microcrystals. The second factor is the large increase of the density of the grain boundaries with a decrease of the average grain size and, correspondingly, the change of the dislocation hardening mechanism to the grain-boundary mechanism. The third factor – with the refining of the grain size the dislocation deformation processes is replaced by grain boundary sliding and this is followed by the increase of the importance of the role of the free volume in deformation. At small grain sizes (ultrafine-grained or nanograins) a significant role is played by differences in the behaviour of the very large and very small grains in the size distribution function for the specific specimen. The largest grains are characterised by sliding at the grain boundaries and by intragranular dislocation activity, including martensitic transformation and twinning dislocations. On the other hand, the finest grains behave as solid non-deformed particles [18, 59].

The generalised pattern of the dominant deformation mechanisms of copper at room temperature is shown in Figure 1.11. The figure shows the size range from 1 nm to 10 μm, including the mean grain sizes from 2.5 nm to the size larger than 10 μm. On the scale, the diagram starts at 1 nm which corresponds to the thickness of the grain boundary interlayers. The interlayers with the size of 1–2.5 nm characterise the start of formation of the solid-state structure. At small sizes, the amorphous structure is present in the initial stage and this is followed by the formation of a crystal structure and formation of the grains. At 3–5 nm the material usually contains the equilibrium crystal structure with the formation of polycrystalline formations containing the general and special type boundaries [19].

The upper part of the diagram in Fig. 1.11 shows the grain structure: up to 100 nm there are dislocation-free grains, at $d > 100$ nm the grains retain the dislocations. Although at $d > 100$ nm the researcher does not find the dislocations in the grains, the dislocation mechanisms operate as shown in the diagram. The upper part of the diagram shows the ranges of the positive and negative values of the Hall–Petch coefficient k [60, 61]. As is well-known, this ratio has the form

$$\sigma = \sigma_0 + kd^{-1/2} \tag{1.4}$$

where σ is the deformation stress of the polycrystal, σ_0 is the deformation stress of the single crystal. The diagram in Fig. 1.11 shows clearly that at $d > 200$ nm the dislocation slip mechanisms are dominant, and at $d < 25$ nm the grain boundary diffusion mechanisms are controlling. In the range 10 nm–1 μm the grain boundary and dislocation mechanisms overlap. The term 'dominant mechanisms' in the caption of Fig. 1.9 indicates that in this grain size range the contribution of these mechanisms to deformation is significant but some of them can also operate outside the given grain size range. These problems are qualified in greater detail in the studies of the authors of this book [18, 19, 60–62].

The change of the work hardening parameters with a change of the grain size is closely associated with the evolution of the deformation mechanisms in transition through the critical parameters of the grain structure. The critical values of the average grain size in the micro- and nanoregions are 10–25, 100 and 1000 nm. This problem will be discussed in greater detail in the chapters 2 and 3. At very small grain sizes the dominant mechanisms are the displacement of the free volume, diffusion along triple lines and the grain boundaries. With the increase of the grain size the lattice diffusion and the sliding of partial dislocations start to operate. With a further increase of the grain size the sliding of the total lattice dislocations, dynamic crystallisation and diffusion along dislocation pipes become dominant.

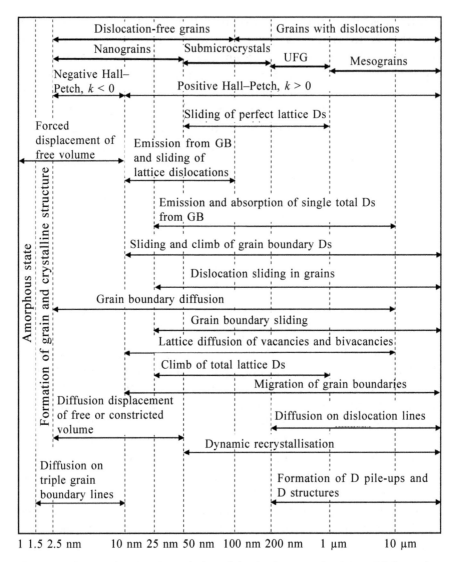

Fig. 1.11. Diagram showing the variation of the dominant mechanisms of deformation for copper with the change of the grain size at $T = 300$ K; D – dislocations, k – the Hall–Petch parameter, GB – grain boundaries.

1.14. Problem of determination of the grain size at the microlevel

In the construction of relationships in the micro- and, in particular, the nanorange in relation to the average grain size it is quite difficult to compare the results obtained by different authors. If all investigators measured the parameters of the grain structure after etching out the grain boundaries in the mesorange by optical microscopy, in the micro-range where the optical measurements cannot resolve the structure, different authors used different methods.

Firstly, it is X-ray diffraction analysis. In most cases, the results of measurements of the diffraction lines are used and this gives the size of the coherent scattering regions (CSR). In another case, measurements are taken of the number of spots on a Debye ring. In the final analysis, the dimensions of the CSR are also estimated. Naturally, the grain size and the size of the CSR are not the same, and the comparison of the results of X-ray diffraction analysis and electron microscopy is not always satisfactory [63, 64].

Secondly, it is scanning electron microscopy (SEM). This method can be used to determine the large misorientations and even measure them. However, the application in most cases of the single-reflection method does not provide a complete pattern of misorientation and the value of the grain size is too large, and in the case of small misorientations the SEM method 'confuses' the misorientation of the grains and blocks.

Thirdly, when using transmission electron microscopy (TEM) and mass determination of the grain sizes, the parameters of the grain sizes are usually not identified, and many investigators use simply the change of the contrast through some boundary.

It may be concluded that the different methods measure different characteristics and all these results are referred to as the 'average grain size'. When reading any article it is not possible to repeat the measurements and it is necessary to use the data obtained by the authors of these investigations. These data are then used for analysis carried out in the present work.

1.15. Identification of plastic deformation stages at the microlevel

In this section attention is given to the examples of identification of the plastic deformation stages based on the analysis of the $\sigma = f(\varepsilon)$ relationships. The first stage is the differentiation of the $\sigma = f(\varepsilon)$ relationship for determining the work hardening coefficient: $\theta(\varepsilon) = d\sigma(\varepsilon)/d\varepsilon$.

Attention will be given to the relationships of plastic deformation of copper polycrystals at the microlevel. The first example – copper polycrystals with the grain size of 20 nm [65]. Figure 1.12a shows the $\sigma(\varepsilon)$ and $\theta(\varepsilon)$ dependences. The $\sigma(\varepsilon)$ dependences already show clearly that there are two linear stages, II and IV, and the transitional quasi-parabolic stage III. The dependence $\theta(\varepsilon)$ confirms the three-stage form of the $\sigma(\varepsilon)$ dependence. This case of the individual stages corresponds to the scheme shown in Fig. 1.10.

The second example – a copper polycrystal with a grain size of 27 nm [66], Fig. 1.12b. The grain size is slightly larger than in the previous case, and the dependence $\sigma(\varepsilon)$ is slightly lower than that in Fig. 1.12a. The stages on the $\sigma(\varepsilon)$ dependence in Fig. 1.12b are less clear but the differentiation with respect to ε indicates the existence of all stages, starting with the stage II and ending with the stage VIII. In this case, the diagram corresponds to Fig. 1.1. In Fig. 1.12b after differentiation the $\theta(\varepsilon)$ dependence demonstrates the stage II, the short stages III, IV and VII and the long stages VI and VIII. In this case $\theta_{II} > \theta_{IV} > 0$, $\theta_{VI} = 0$ and $\theta_{VIII} < 0$.

The next part of the section is based on the application of the experimental data, in particular, the relationships $\sigma(\varepsilon)$, taken from [66–71]. The data obtained in these studies were used for the same investigations as the data in Fig. 1.12a, b. The calculated dependences $\theta(\varepsilon)$ were used to identify the deformation stages for approximately 20 copper specimens and to evaluate the parameters of the stages, in particular, $\theta(\varepsilon)$, for the even stages.

The currently available set of the experimental data is quite small and it is not possible to determine the reasons for which only one

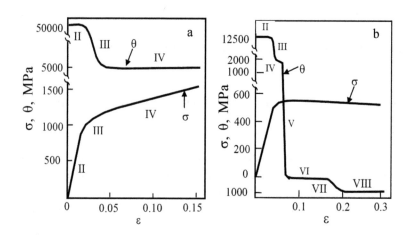

Fig. 1.12. The $\sigma = f(\varepsilon)$ and $\theta = f(\varepsilon)$ dependences for two nanopolycrystals: a – $d = 20$ nm, b – $d = 27$ nm. The dependences show the individual deformation stages.

of the possible stages IV, VI or VIII or even two of these stages successively, take place. It may only be confirmed that the stage IV is present on the $\sigma = f(\varepsilon)$ dependence more frequently, and the stages VI and VIII – less frequently. The following combinations are found in some cases: IV–VI and less frequently IV–VI–VIII. In the latter case, the diagram in Fig. 1.10 is replaced by Fig. 1.11. The presence of the stage VI and, in particular, the stage VIII is associated with the localisation of deformation and may become evident especially in cases in which the grain-boundary hardening is accompanied by a significant contribution of the intragranular types of hardening: solid solution, dislocation and/or dispersion.

1.16. The stress σ–strain ε dependence for copper polycrystals with different nanograin sizes

The study [67] contains the data on the effect of the average grain size on the form of the $\sigma = f(\varepsilon)$ flow curves for the grain sizes of 26, 49, 110 nm and 20 μm (Fig. 1.13) in the tensile test of the nanocrystalline specimens produced by the method of inert gas condensation followed by thermal compacting. The specimens were tensile loaded. Figure 1.13 shows that the refining of the grains changes greatly the characteristics of the deformation stages – the duration of stage III increases, the rate of decrease of the work hardening coefficient decreases, the stress σ_{VI} in stage VI (with constant θ) increases, and ductility decreases.

Figures 1.14–1.16 shows the $\sigma = f(\varepsilon)$ relationships for compressive strain [65, 66, 68, 69]. As in compression, tensile loading is also accompanied by the occurrence of stage III with decreasing work hardening coefficient $\theta = d\sigma/d\varepsilon$ for all grain sizes. This is followed by the long linear hardening stage. It is stage IV, if θ > 0 (Fig. 1.14, curves 1, 2, 4, 5), or stage VI, if θ = 0, or even stage VIII if θ <0 (Fig. 1.14, curve 3). With a decrease of the average grain size the stress in the stages IV, VI and VIII increases. Stage IV is replaced by the stages VI–VIII (Figs. 1.14 and 1.15 [69]). However, the data obtained in [66, 68, 69] show that a decrease of the grain size does not always increase the flow stress. This is clear from the comparison of the curves for the nanocrystals with the grain sizes of 25 and 27 nm (curves 3 in Figs. 1.14 and 1.15), and also from the comparison of the flow stresses for the nanocrystals with the grain size of 50 and 54 nm (Fig. 1.14, curve 5 and Fig. 1.15, curve 2). In both cases, the flow stress differs by at least 200 MPa.

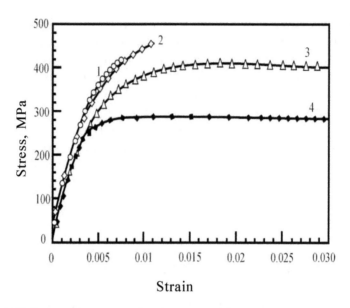

Fig. 1.13. Deformation curves for the copper polycrystals [67] with the grain size of 26 nm (1), 49 nm (2), 110 nm (3) and 20 μm (4).

The data in [65] (Fig. 1.16) differ by the large scatter of ductility values and the extremely high acting stresses which are 2–3 times higher than the values obtained in [66, 68, 69] (Figs. 1.14–1.16). It would be useful to compare them with the flow stresses for the stage VI for the copper single crystals with the average grain size of 200 nm. The compression tests of a compacted copper specimen [62] (Fig. 1.17) show that the flow stress for the specimen with a grain size of 200 nm is twice as high as for the specimen with a grain size of 25–50 nm.

1.17. Relationships of work hardening of copper micropolycrystals with different grain sizes

The analysis of the current state of the problem of the work hardening of copper micropolycrystals has been used to determine a number relationships and associated problems. Firstly, it is the purity of the material. Nanocopper often contains oxygen or phases with oxygen. In the latter case, the nanocopper is hardened in comparison with pure copper and may be embrittled. The higher deformation resistance of the copper polycrystals with the grain size of 20 nm (Fig. 1.16) in comparison with the copper polycrystals with the grain size of 27 nm (Fig. 1.14b) without extensive contamination with the

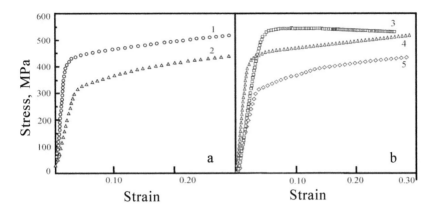

Fig. 1.14. Deformation curves for the copper nanocrystals with the grain size of 54 (1), 82 (2), 27 (3), 39 (4) and 54 nm (5): a – the data from [68]; b – the data from [66].

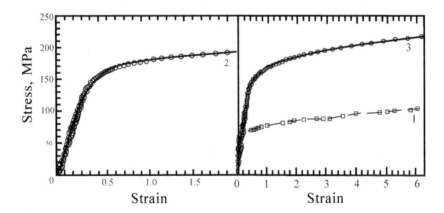

Fig. 1.15. Deformation curves for the copper nanocrystals with the grain size of 50 μm (1), 50 nm (2) and 25 nm (3) (strain rate $1.4 \cdot 10^{-5}$, s^{-1}) [69].

impurities cannot be explained. Secondly, after compacting, the copper specimens are characterised by reduced density and contain micropores. Consequently, the specimen is softened and can fail by brittle fracture. Thirdly, the problem of measuring the average grain size in the presence of fine grains has not been completely solved. Objective data can be obtained only using transmission electron microscopy with the determination of the parameters of the boundaries. If the parameters of the boundaries are not determined, the data can be too low because the dimensions of the grains and sub-grains are averaged-out. When determining the grain size by

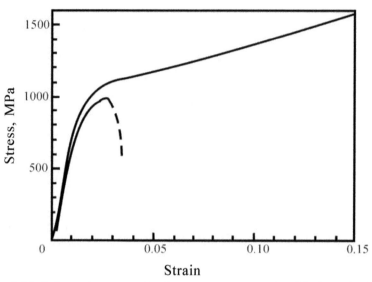

Fig. 1.16. Deformation curves for the copper nanocrystals with the grain size of 20 nm (strain rate 10^{-3}, s^{-1}) [65].

x-ray diffraction analysis which provides information on the size of the coherent scattering blocks, there is also a risk of obtaining too low data for the grain size. To verify that the average grain sizes in many published studies were too low, one can use the method based on the analysis of the flow stress at the beginning of the stages IV, VI and VIII. The flow stress at the beginning of the stage VIII should increase with the reduction of the grain size, at least up to the moment when the grain boundary sliding provides a larger contribution to the deformation than the intragranular dislocation sliding. Comparison of the data on the deformation resistance of the copper specimens with the average size of 27 to 82 nm (Fig. 1.14) and 200 nm (Fig. 1.70) shows that the data for the average grain size, obtained by x-ray diffraction analysis and electron microscopy, differ.

The literature data show that there are other factors affecting the form of the $\sigma = f(\varepsilon)$ dependence. They include test temperature, strain rate, the size of the sample or its relationship with the grain size. It is also important to take into account the technological conditions, such as annealing after preparation of the sample (whether annealing is carried out or not).

Several of these relationships will now be discussed. Firstly, the fine- and nanocrystalline specimens show after stage III stages with a constant hardening coefficient θ: at $\theta > 0$ it is stage IV; at $\theta = 0$ stage VI and at $\theta < 0$ stage VIII. In these stages the Backofen–

Konsider condition is usually satisfied [6]. The coordinate at which $\sigma = \theta$ on the $\sigma = f(\varepsilon)$ dependence is shown in Fig. 1.17. When this condition is fulfilled the stage VI or similar stages VII or VIII start. These stages are characterised by the development of localisation of deformation. This phenomenon is especially distinctive in stage VIII [61], Fig. 1.18. In this case the dependence $\sigma = f(\varepsilon)$ is characterised by the maximum σ_{max} at some strain denoted by ε_{max}. Above this maximum deformation localisation takes place.

A decrease of the average grain size is accompanied by the following tendencies:

1) the beginning of the stages IV–VIII takes place at higher flow stresses;

2) the probability of the change of stage IV to the stages VI or VIII becomes higher (Fig. 1.10);

3) the ductility decreases, and the transition to the stages IV–VIII may not take place at all as a result of fracture;

4) the localisation of deformation leads to σ_{max} with a subsequent decrease of the flow stress (Fig, 1.16 [72]). As the grain size decreases, the value of σ_{max} increases and the strain ε_{max} at which σ_{max} is reached decreases.

The current state of the problem of the work hardening of the polycrystalline pure metals with the average grain size in the nanorange indicates that the investigations of these problems cannot be regarded as completed. The main problem is to obtain the accurate characteristics of the structural state of the investigated polycrystals – the grain and subgrains size, the grain boundary and sub-boundary structure, the density of the statistically stored and geometrically necessary dislocations; internal stress fields, the impurity concentration, the presence of secondary phases, disruption of the integrity and spatial localisation of the particles of the secondary phases, impurities and discontinuities.

At the same time, the experimental results have been used to reveal a number of relationships of the development of the individual stages of the work hardening of nanopolycrystals and to show the possibility of obtaining high values of deformation resistance. For pure copper with the average grain size of 20–200 nm these values are in the range 500...1000 MPa at a considerably high ductility. In this case, it is important to explain quantitatively the contributions to the deformation of the local slip mechanisms characteristic of the nanorange [60, 62]. To solve this problem, it is important to

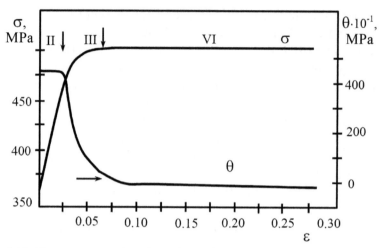

Fig. 1.17. The stress σ–strain ε dependence for copper with the mean grain size of 20 nm (horizontal arrow – the coordinate at which the Backofen–Konsider condition is satisfied; vertical arrows – the boundaries of the deformation stages).

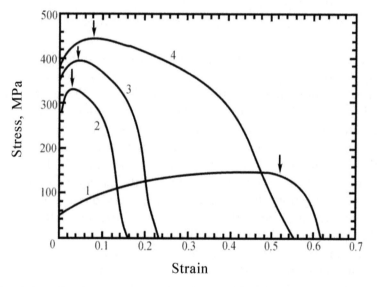

Strain

Fig. 1.18. Deformation curves for copper deformed at 293°C (strain rate 10^{-3} s^{-1}); 1 – coarse-grained copper, 2 – copper after cold rolling (60%), 3 – nanocopper (equal channel angular pressing, 2 passes), 4 – nanocopper (equal channel angular pressing, 16 passes). The arrows indicate σ_{max} and ε_{max} [72].

investigate in considerable detail the evolution of the structure of the nanopolycrystalline aggregate during deformation.

1.18. Hardening mechanisms and special features of the individual stages of deformation of polycrystals with nanograins

The problem, described in the initial section of the chapter, is important and interesting. The ductility of polycrystals with the nanograins is directly associated with the relationships governing work hardening and evolution of the defective structure, determined by plastic deformation [18]. A special pattern of the individual stages of deformation of the nanograins polycrystals with the retention of only a part of the stages observed in the coarse-grained crystals is important for understanding the ductility margin owing to the fact that the nanopolycrystals are characterised by the fulfilment of the Backofen–Konsider relationship [6]. This condition, expressed by the equality of the values of the flow stress (σ) and the work hardening coefficient (θ), $\sigma = \theta = d\sigma/d\varepsilon$, determines the formation of distinctive localisation of deformation, the possible development of the neck and creates a suitable situation for subsequent failure of the material.

At present, the relationships governing the work hardening of the crystalline aggregates have been investigated mostly on pure metals with a specific grain size. The results obtained for the FCC pure metals show that in the range of the nanograins the dependences σ–ε show the short stage III which may also be preceded by the short stage II. The stages are usually replaced by the long stage with linear hardening. This may be either the stage IV, if the work hardening coefficient is $\theta = d\sigma/d\varepsilon > 0$, or stage VI, if $\theta = 0$, the long stage with a decreasing value of $\theta < 0$, i.e., the stage VIII (Fig. 1.10).

If in the case of the pure metals with the nanograins the data for work hardening are insufficient and it is difficult to investigate the dependence of the work hardening parameter of the grain size, for the alloys these data is extremely scarce. This and the following sections examine the transition from the pure metals to the alloys and its influence on the pattern of work hardening. The pure metals are represented by nanocrystalline copper and gold. Their behaviour under work hardening is compared with the behaviour of the nanocrystalline alloys based on copper at comparable nanograin sizes. The copper-based alloys can be processed by solid solution, dispersion and polyphase hardening.

Initially, the work hardening of nanocrystalline copper is compared with nanocrystalline gold. Both nanocrystalline copper and gold may show grain boundary and dislocation hardening. The σ–ε relationship for these metals is shown in Fig, 1.19 [66, 67]. The average grain size of copper is 40 nm, and that of gold 36 nm. The ductility of gold is restricted to several percent of deformation, and the copper is characterised by higher ductility. The σ–ε dependences to a strain of ε = 0.02 are almost identical indicating, firstly, that the results are highly reliable and, secondly, that this agreement occurs at almost equal dimensions of the nanograins of copper and gold. The form of the σ–ε dependences in Fig. 1.19 indicates the presence of the stages II and III and cannot be used to analyse further stages. Attention should be given to the high flow stress reaching 400 MPa.

1.19. Effect of different hardening mechanisms on the flow stress and the form of the σ = f (ε) dependence

Figure 1.20 shows the comparison of the σ–ε relationships for nanocrystalline copper [60] and the Cu +0.5 wt.% Al_2O_3 copper alloy [73]. The average grain size of copper is 82 nm, in the copper alloy 80 nm, i.e., the grain sizes of these two materials are almost identical. As indicated by Fig. 1.20, the stages II and III of work hardening are detected in the initial phases. At a strain higher than 6...7% the long-term stages with linear hardening were formed. In nanocrystalline copper it is the stage IV (θ > 0), and this stage starts at

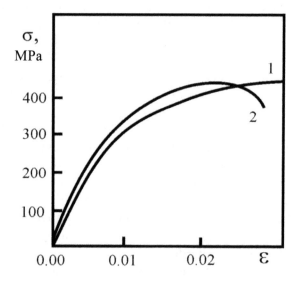

Fig. 1.19. The σ–ε dependence for copper nanopolycrystals (1) with the grain size of 40 nm [66] and gold nanopolycrystals (2) with a grain size of 36 nm (2) [71].

a stress of σ_{IV} = 350 MPa. In the nanocrystalline dispersion-hardened alloy it is the stage VI (θ = 0). This stage starts at a stress of σ_{VI} ≈ 630 MPa. The contribution of dispersion hardening at exit to the linear stage of hardening and different values of the flow stress of the dispersion hardened alloy in stage VI and copper in stage IV (630 MPa–350 MPa = 280 MPa) is large.

Attention will be given to the special main features of work hardening: 1) the linear stage of the dispersion-hardened copper alloys starts at stresses almost twice as high as in pure copper; 2) the ductility of the dispersion-hardened alloy is considerably lower than that of pure copper, in the alloy after ε ≈ 0.18 there is strong localisation of deformation and neck formation. The higher degree of hardening of the copper alloy in the linear stage indicates the possible significant contribution of dispersion hardening of the polycrystals in the nanorange. In particular, in the case of curve 2 in Fig. 1.23 the size of the particles of Al_2O_3 in the Cu +0.5 wt.% Al_2O_3 alloy is equal to 10 nm, and the particles are distributed quite uniformly throughout the volume of the material. Comparison of the curves 1 and 2 in Fig. 1.20 indicates the important role of intragranular deformation processes in this dispersion-hardened alloy.

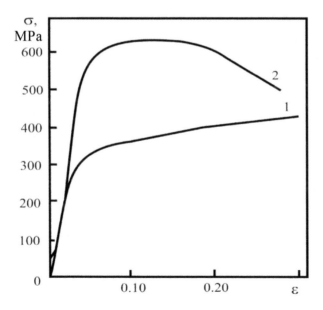

Fig. 1.20. Comparison of the σ–ε dependences for nanopolycrystals of pure copper (1) with a grain size of 82 nm [68] and the dispersion-hardened nanopolycrystalline copper alloy Cu +0.5 wt.% Al_2O_3 (2) with a grain size of 80 nm [73].

Figure 1.21 compares the σ–ε relationships for the nano-polycrystals of pure copper (curve 1) [67], the Cu+0.44 wt.% Cr +0.2 wt.% Zr solid solution and the dispersion-hardened alloy with the same composition [70]. The average grain size of pure copper is 110 nm, in the solid solution it is 160 nm, and in the dispersion-hardened alloy it is also 160 nm, the chromium particle size is 12 nm. The grain sizes for the relationships 1, 2 and 3 are similar and, therefore, it is possible to separate the contributions of solid solution and dispersion hardening. In most cases, the particles are localised at the grain boundaries. All the three relationships σ = $f(\varepsilon)$ in Figure 1.21 show the short stages II and III, replaced by the long stages VI ($\theta = 0$, curve 1), IV ($\theta > 0$, curve 2) and VIII ($\theta < 0$, curve 3). The beginning of the long stages IV, VI, and VIII is found at the following stresses: $\sigma_{VI} = 400$ MPa, $\sigma_{IV} = 440$ MPa, $\sigma_{VIII} = 640$ MPa. The behaviour of pure copper, corresponding only to the dislocation hardening, is expressed by the value $\sigma_{VI} = 400$ MPa. When the solid solution hardening provides a contribution (curve 2), the linear stage starts at a higher stress $\sigma_{IV} = 440$ MPa. The contribution of solid solution hardening is not large but quite visible, $\sigma_{IV} - \sigma_{VI}$ (440 MPa–400 MPa = 40 MPa). Since the grain size of copper is slightly smaller (110 nm) in comparison with the

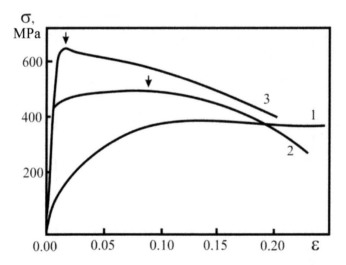

Fig. 1.21. The σ–ε dependences for the nanopolycrystals of copper (d = 110 nm) (1) [67], the Cu + 0.44 Cr +0.2 Zr (wt.%) solid solution (2) and the aged solid solution with the chromium precipitates with the average size of 12 nm (3) [70]. In the alloy d = 160 nm. The arrows indicate the values of σ at which the Backofen–Konsider condition σ = θ is satisfied and macroscale localisation of deformation starts to take place.

alloy (160 nm), the solid solution hardening is actually greater than 40 MPa. The contribution of dispersion hardening $\Delta\sigma_{disp}$ = $\sigma_{VIII} - \sigma_{VI}$ (640 MPa – 400 MPa = 240 MPa) is large. This is natural because the dispersion hardening is usually more pronounced than the solid solution hardening. It is possible to compare the values of the dispersion hardening for the relationships shown in Figs. 1.20 and 1.21. The value $\Delta\sigma_{disp}$ = 280 MPa in Fig. 1.20 and $\Delta\sigma_{disp}$ = 240 MPa in Fig. 1.21. Firstly, it is larger in value and comparable with the contribution of work hardening. Secondly, for the cases shown in Figs. 1.20 and 1.21, the resultant values are similar. Thirdly, the contribution of dispersion hardening decreases with the increase of the average grain size (in Fig. 1.20 $<d>$ = 80 nm, in Fig. 1.21 $<d>$ = 110–160 nm).

Figure 1.22 compares the σ–ε relationships for nanopolycrystalline pure copper with the average grain size of $<d>$ = 210 nm and the Cu–Al–O alloy based on copper with the average grain size of $<d>$ = 200 nm [18, 62, 74]. In contrast to the copper polycrystals with the average grain size of $d < 100$ nm, at the grain size $d > 100$ nm the former are characterised by the formation of a dislocation substructure, either cellular or network [18, 74]. The grain, grain-boundary and dislocation structures develop in the course of plastic deformation at $d > 100$ nm. At $d < 100$ nm the processes of evolution of the grain structure also take place but this process has not been studied sufficiently [75]. The grain boundaries and, in particular, the contacts of the grain boundaries in nanocrystalline copper contain second phase particles, Cu_3Sn (Sb), Cu_3N, Cu_2O and CuO. The particle size varies from several tens of nanometres to several nanometres (the size decreases in the order of sequence of these phases) [76, 77]. The formation of the secondary phases is determined by the methods of preparation of the materials. The mean grain size of the Cu–Al–O alloy is $d > 200$ nm. The alloy contains grains of different type – copper grains, the grains of the solid solution of aluminium in copper and two-phase and polyphase grains based on copper, containing particles of Cu_9Al_4, $CuAl_2$, α-Al_2O_3 and Cu_2O. The size of these particles is 25...200 nm. The individual particles with the size from 100 nm to 1 μm were classified as secondary grains. In addition to this, the grain boundaries and, in particular, the junctions of the grains contained fine particles of Cu_2O and α-Al_2O_3. The size of these particles was in the range 5...50 nm. Thus, both the copper specimens and the specimens of the Cu–Al–O alloy were hardened by the effect of different mechanisms,

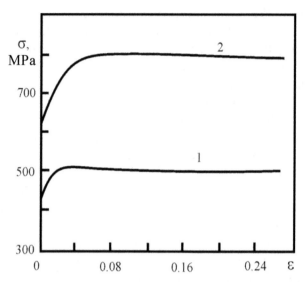

Fig. 1.22. The stress σ – true strain ε dependence for the fine-grained materials: Cu (1) and Cu–Al–O alloy (2). The average grain size $<d>$ = 210 nm for Cu and $<d>$ = 200 nm for the Cu–Al–O alloy.

in addition to grain boundary hardening. In the case of copper specimens it was mostly intergranular substructural hardening, in the case of the alloy – mostly polyphase hardening in the volume and dispersion hardening at the grain boundaries.

Attention will now be given to the individual stages of the σ = $f(ε)$ relationships and the hardening parameters shown in Fig. 1.22 for the ultrafine copper and Cu–Al–O alloy [18]. Firstly, both curves in Fig. 1.22 contains the short stages II and III which are replaced by the long stage VI in which θ = 0. For copper $σ_{VI}$ = 500 MPa, for the Cu–Al–O alloy $σ_{VI}$ = 800 MPa. The increase of the flow stress for the copper polycrystal in Fig. 1.22 in comparison with the data shown in Figs. 1.19–1.21, which took place regardless of the increase of the grain size, is obviously associated with the substructural hardening in the body of the grains of the copper polycrystals. Owing to the fact that the value of $σ_{IV}$ or $σ_{VI}$ for the small grain sizes does not reach 400 MPa, whereas for $<d>$ = 210 nm, the value $σ_{VI}$ = 500 MPa, then at least 100...150 MPa can be related to the substructural hardening ($Δσ_{sub}$) in the body of the copper grains. It is well-known that $<d>$ = 100 nm is the critical grain size for the transition from dislocation-free grains to the grains with a dislocation substructure [18, 60, 62]. The intragranular hardening mechanisms change in this case.

The polyphase hardening, equal to the difference of the flow stress in the stage VI of the copper alloy and copper, reaches $\Delta\sigma_{ph} =$ 300 MPa and is slightly higher than the value of dispersion hardening ($\Delta\sigma_{disp}$), Figs. 1.20 and 1.21, which varied in the range 240–280 MPa.

1.20. Basic pattern of work hardening of nanocrystals

In the previous sections, the behaviour during work hardening of the nanocrystals of gold and copper and nanocrystals of copper-based alloys was compared. The experiments were carried out using specimens with similar grain sizes in the range 36...200 nm. The dependences $\sigma = f(\varepsilon)$, formed during the operation of different hardening mechanisms, were examined. These mechanisms include: grain boundary barrier hardening (1) and dislocation substructural hardening (2), taking place in both pure copper and all the examined nanopolycrystals. The Cu–Al–O alloy (Figs. 1.20 and 1.22) also showed the mechanisms such as dispersion hardening (6) and inhibition of grain boundary sliding (4). The Cu–Cr–Zr alloy (Fig. 1.21) was characterised by solid solution (5) and dispersion hardening (3). Finally, the Cu–Al–O alloy (Fig. 22) showed polyphase hardening (6) in addition to dispersion hardening (3).

Regardless of the abundance of the acting hardening mechanisms (six mechanisms), the scheme of work hardening stages, previously obtained for the pure nanometals and usually consisting of at least three different stages [61] remains unchanged. For the FCC nanopolycrystals investigated previously with the average grain size of 10...250 nm [61] examination showed the following stages (Fig. 1.23). Firstly, it is the stage II with a linear high constant value of θ. This stage is followed by stage III with a decreasing value of θ. Work hardening is completed by one of the possible stages with a constant value θ. It can be the stage IV ($0 > 0$), or stage VI ($\theta = 0$), or stage VIII ($\theta < 0$).

The first critical studies of the analysis of the relationships of work hardening of the materials with a grain size of 12...210 nm were carried out by the authors of this book [60, 61, 78]. These studies analysed a large number of $\sigma = f(\varepsilon)$ the relationships, obtained on materials such as copper, cobalt, copper and aluminium alloys in the grain size range 12...230 nm. The results obtained for these materials show that at these grain sizes the $\sigma = f(\varepsilon)$ relationships usually show three stages, namely: stages II, III and the final long stage, which can be either stage IV, VI or VIII. Consequently, the systematic representation of different variants of the pattern of work hardening (softening) is the one shown in Fig. 1.23.

If the grain boundary hardening is accompanied by a contribution from other hardening mechanisms (dislocation, solid solution, dispersion, multiphase) and if the contributions of these mechanisms are large (or the contribution of at least one of them), the flow curves show rapidly the implementation of the Backofen–Konsider condition [6], with the formation of the zone of deformation localisation – the neck in tensile loading or 'shear' in compression, with the start of the stage with a negative work hardening coefficient. It may be concluded that the inclusion in the nanomaterials of other hardening mechanisms with a large contribution in addition to grain boundary hardening greatly changes the flow curve, as shown in Figs. 1.20–1.22. This indicates that the transition to the nanometric grains undoubtedly decreases the plasticity margin and, therefore, the grain boundary hardening may be accompanied by other types of hardening. The latter fact makes it possible to obtain high strength values of the nanopolycrystals.

1.21. Effect of the grain size on the parameters of plastic deformation stages

The main characteristics of the plastic flow stages are, firstly, the work hardening coefficients $\theta = d\sigma/d\varepsilon$, secondly, the duration of the stages and, thirdly, the stress of the beginning and end of the stages. The most important characteristic is the work hardening coefficient. Its dependence for the main even stages II, IV and VI on the grain size of copper, deformed at room temperature, is shown in Fig. 1.24. To obtain this relationship, the authors processed a large number of $\sigma = f(\varepsilon)$ graphs, obtained by different authors. Regardless of the large scatter determined by the individual methods of preparation of the copper specimens with different grain sizes in the range 20...200 nm, there is a certain dependence of θ in different deformation stages on the grain size.

It may be seen that θ_{II} (Fig. 1.24a) reaches extremely high values in the vicinity $d = 20...40$ nm. The value θ_{II} rapidly decreases with increasing grain size. At $d > 50$ nm the rate of decrease of θ_{II} decreases, and at $d > 200$ nm the value θ_{II} approaches the value θ_{II} for the polycrystals with the grain size at the mesolevel. The value θ_{IV} is much smaller than θ_{II} but the behaviour of $\theta_{IV} = f(<d>)$ is very similar to that of θ_{II}, and θ_{IV} reaches the maximum value close to

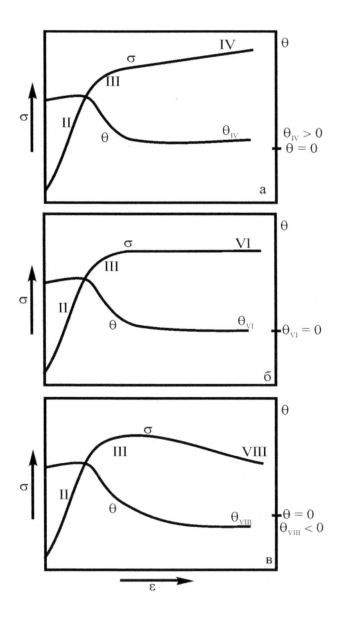

Fig. 1.23. Diagram of three different variants of the chart of work hardening (softening) at different strains of the FCC nanopolycrystals.

$<d>$ = 20 nm and then rapidly decreases with increasing $<d>$, and at $d > 50...100$ nm becomes similar to the appropriate parameter for the grains at the mesolevel.

According to the definition, θ_{VI} is equal to 0 and in accordance with the measurement results remains unchanged in the entire examined grain size range, including at $<d>$ = 25...50 nm. Figures 1.24a, 1.12b and 1.11 show that the maximum values of θ_{II} and θ_{IV} are obtained at the grain sizes $<d>$ corresponding to the boundary of grain boundary – the diffusion deformation mechanism, on the one side, and lattice–diffusion and grain boundary–dislocation deformation mechanisms, on the other side (Fig. 1.11). The transition to the lattice–dislocation mechanisms indicates that the values of θ_{II} and θ_{IV} at the microlevel approach the values of these parameters for the mesolevel.

The above mentioned graphs indicate that the work hardening parameters of the polycrystals at the microlevel change in relation to the grain size in strict correspondence with the change of the dominant deformation mechanisms. With increasing grain size these mechanisms change from grain boundary–diffusion to lattice–diffusion and grain boundary–dislocation and then to lattice-dislocation. Although the well-known Kim–Estrin–Bush [79, 80] and Conrad [81] theoretical schemes do not take into account all the deformation mechanisms shown in Figure 1.11, nevertheless, these schemes correspond to the conclusions described in this section.

It may be concluded that the multistage $\sigma = f(\varepsilon)$ dependence, characteristic of the polycrystals at the mesolevel, remains valid to a large extent also for the microlevel and the nanorange of the grain sizes. The deformation curves always contains the short stages II and III and they are followed by the long stages IV or VI or VIII. In the nanosized grain range the work hardening coefficients θ_{II} and θ_{IV} reach the maximum values. With increasing nanograin size the values θ_{II} and θ_{IU} rapidly decreases and in the range 100–200 nm reached the values characteristic of the grain size at the mesolevel.

The variation of the work hardening parameters with the variation of the grain size is closely associated with the evolution of the deformation mechanisms in transition through the critical parameters of the grain structure. The critical mean grain sizes are 10, 25, 100 and 1000 nm. At very small grains the dominant mechanisms are the displacement of the free volume, diffusion at triple lines and grain boundaries. An increase of the grain size results in the activation of lattice diffusion and the gliding of partial dislocations. With a further

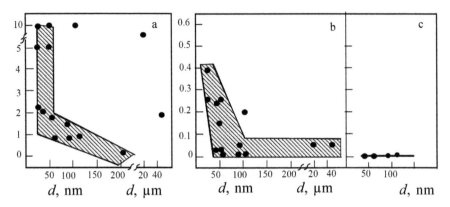

Fig. 1.24. Dependence of the work hardening coefficient θ of the even stages on the average grain size d of copper: a) θ_{II}, b) θ_{IV} and c) θ_{VI}.

increase of the grain size the slip of the total lattice dislocations, dynamic crystallisation and diffusion along the dislocation pipes become dominant, and the dislocation substructure starts to form in the body of the grain and the glide systems with different Schmid factors start to operate.

References

1. Bell J.F., Experimental foundations of the mechanics of deformable solids. Part I. Small strains. (Trans. from English. Ed. A.P. Filin). Moscow: Nauka, 1984. 596 pp.
2. Bell J.F., Experimental foundations of the mechanics of deformable solids. Part II. Finite strains. (Trans. from English. Ed. A.P. Filin). Moscow: Nauka, 1984.
3. Trefilov V.I., et al. Work hardening and fracture of polycrystalline metals. (Ed. V.I. Trefilov), Kiev: Naukova Dumka, 1989.
4. Panin V.E., et al., The structural levels of plastic deformation and fracture. (Ed. V.E. Panin), 1990.
5. Koneva N.A., Kozlov E.V., *Izv. VUZ Fizika,* 2004, No. 8, 90–98.
6. Backhofen V., Deformation processes, Moscow, Metallurgiya, 1977.
7. Friedman Ya.B., Mechanical properties of metals. Volume 1. Deformation and destruction, Moscow, Mashinostroenie, 1974.
8. Sachs G., Weerts I., *Z. Phys.,* 1930, V. 62, 473–481.
9. Koneva N.A., Kozlov E.V., The physical nature of stages of plastic deformation, in: Structural levels of plastic deformation and fracture (Ed. V.E. Panin), Nauka, Novosibirsk, Siberian Branch, 1990, 123–186.
10. Zeeger A., The sliding mechanism and strengthening in the face-centered cubic and hexagonal close-packed metals, in: Dislocations and mechanical properties of crystals. Moscow, IL, 1960, 179–289.
11. Nes E., *Progr. in Mater. Sci.,* 1998, V. 41, 129–193.
12. Koneva N.A., et al., Parameters of dislocation structures and factors determining flow stress at stages III and IV, in: Strength of metals and alloys. Proceed. of ICSMA-8, Oxford, Pergamon Press, 1998, 385–390.

13. Ultrafine grains in metals. (ed. L.K. Gordienko), Moscow, Metallurgiya, 1973.
14. Valiev R.Z., Alexandrov I.V., Nanostructured materials produced by severe plastic deformation, Moscow, Logos, 2000.
15. Severe Plastic Deformation, New York, Nova Science Publishers, 2006.
16. Meyers M.A., et al., *Progr. Mat. Sci.*, 2006, V. 51, 427–556.
17. Kozlov E.V., et al., in: Functional-mechanical properties of materials and computer design, (ed. V.A. Likhachev), Pskov Publishing house ADVELA, 1993, 221–225.
18. Kozlov E.V., et al., *Fiz. mezomekhanika*, 2004, V. 7, No. 4, 93–113.
19. Kozlov E.V., et al., *Fiz. mezomekhanika*, 2006, V. 9. No. 3, 81–92.
20. Kozlov E.V., et al., *Izv. VUZ, Fizika*, 1991, No. 3, 112–128.
21. Kozlov, E.V., et al., *Izv. VUZ, Chernaya metallurgiya*, 1994, No. 8, 35–39.
22. Kozlov E.V., et al., *Izv. VUZ, Fizika*, 2002, No. 3, 49–60.
23. Fang X.F., Dahl W., *Mater. Sci. Eng.,* 1995, V. A 203, 14–25.
24. Fang X.F., Gusek C.O., Dahl W., *Mater. Sci. Eng.*, 1995, V. A 203, 26–35.
25. Fang X.F., Dahl W., *Mater. Sci. Eng.*, 1995, V. A 203, 36–45.
26. Lan Y., Kbaar H.J., Dahl W., *Met. Trans.*, 1992, V. A.23, 537–544.
27. Lan Y., Kbaar H.J., Dahl W., *Metal. Trans.*, 1992, V. A23, 545–548.
28. Firstov S.A., Pechkovskii E.P., *Problemy materialovedeniya*, Number 1, 2002, (29). 70–86.
29. Koneva N.A., et al., *Metallofizika*, 1991, No. 10, 56–70.
30. Konev N.A., Kozlov E.V., *Izv. AN, Ser. Fizicheskaya*, 2002, V. 66. No. 6, 824–829.
31. Kozlov E.V., et al., *Metally*, 1993, No.5, 152–161.
32. Koneva N.A., et al., *Mater. Sci. Eng.*, 1997, V. A234–236. 614–616.
33. Hansen N., Kuhlmann-Wilsdorf D., *Mater. Sci. Eng.*, 1986, V. 81, 141–161.
34. Kuhlmann-Wilsdorf D., *Mater. Sci. Eng.*, 1989, V. A113, 1–41.
35. Nazarov A.A., et al., *Nanostructured materials*, 1994, V. 4, 93–101.
36. Kozlov E.V., Koneva N.A., *Mater. Sci. Eng*, 1997, V. A234–236, 982–985.
37. Kozlov L.A., et al., *Izv. VUZ, Fizika*, 1996, V. 39. No. 3, 33–56.
38. Malygin G.A., *Usp. Fiz. Nauk*, 1999, No. 9, 979–1010.
39. Konev N.A., Kozlov E.V., *Izv. VUZ, Fizika*, 1990, No. 2, 89–106.
40. Konev N.A., *Soros Educational Journal*, 1998, No. 10, 99–105.
41. Konev N.A., Dislocation structures in metals and alloys, Encyclopedia. Modern science. Condensed Matter Physics, Moscow, Magistr Press, 2000, 36–43.
42. Rybin V.V., Large plastic deformation and fracture of metals, Moscow, Metallurgiya, 1986.
43. Rybin V.V., *Izv. VUZ, Fizika*, 1991, V. 34, No. 3, 7–22.
44. Rybin V.V., *Problemy materialovedeniya*, Number 1, 2002, (29), 11–33.
45. Hansen N., Jensen J.S., *Phil. Trans. R. Soc. Lond.* A, 1999, V. 357, 1447–1469.
46. Huang X., et al., *Mater. Sci. Eng.*, 2003, V. A340, 265–271.
47. Romanov A.E., *Solid State Phenomena*, 2002, V. 87, 47–56.
48. Seefeldt M., Klimanek P., *Solid State Phenomena*, 2002, V. 87, 93–112.
49. Gutkin M.Yu., et al., Solid State Phenomena, 2002, V. 87, 113–120.
50. Kozlov E.V., et al., *The Physics of Metals and Metallography*, 2000, V. 90. Suppl. P, 559–567.
51. Koneva N.A., Kozlov E.V., *Izv. RAN, Ser. Fizicheskaya*, 2002, V. 66, No. 6, 824–829.
52. Popova N.A., Lapsker I.A., in: Plastic deformation of alloys, Tomsk, TSU, 1986, 240-248.
53. Teplyakova L.A., et al., in: Alloys with shape memory and other advanced materials.

Part 2, St. Petersburg Research Institute of Mathematics and Mechanics, St. Petersburg State University, 2001, 361–366.

54.   Sharkeev Yu.P., *Fiz. Met. Metalloved.,* 1985, V. 60, Vol. 4, 816–821.

55.   Koneva, N.A.,et al., in: Physics and Solid State Electronics. Interuniversity collection of scientific papers. Issue V, Izhevsk, Publishing House of the USU, 1982, 88–97.

56.   Koneva, N.A., et al., in: Physics of defects of the surface layers of materials (Ed. A.E. Romanov), Leningrad, A.F. Ioffe Institute, 1989, 113–131.

57.   Perevalova O.B., *Fiz. Met. Metalloved.*, 2003, V. 95, No. 6, 85–93.

58.   Kubin L.P., Mortcusen A., *Scr. Mat.,* 2003, V. 48, 119–125.

59.   Kozlov E.V., *Mat. Sci. Eng.*, 2004, V. A387–A389, 789–794.

60.   Kozlov E.V., et al., *Voprosy materialovedeniya*, 2007, No. 4 (52), 156–168.

61.   Kozlov E.V., et al., *Fiz. mezomekhanika,* 2007, V. 10, No. 3, 95–103.

62.   Kozlov E.V., Structure and resistance to deformation of UFG metals and alloys, in: Severe Plastic Deformation (Ed. B.S. Altan), New York, Nova Science Publish., 2006, 295–332.

63.   Koneva N.A., et al., in: Structural and phase states and properties of metallic systems, (Ed. Potekaev A.I.), Tomsk: Publishing house of the NTL, 2004, 83–110

64.   Koneva N.A. Internal long-range stress fields in ultrafine grained materials // Severe Plastic Deformation (Ed. B.S. Altan), New York, Nova Science Publish., 2006, 249–274.

65.   Youngdahl C.J., et al., *Scripta Mat.*, 1997, V. 37, No. 6, 809–813.

66.   Suryanarayanan R., et al., *Mat. Sci. Eng.*, 1999, V. A264, 210–214.

67.   Sanders P.G., et al., *Acta mater.,* 1997, V. 45, No. 10, 4019–4025.

68.   Suryanarayanan R., et al., *J. Mat. Res. Society*, 1996, V. 11, No. 2, 439–446.

69.   Neiman G.W., et al., *J. Mat. Res. Society*, 1991, V. 6, No. 5, 1012–1027.

70.   Vinogradov A., et al., *Acta. Mat.*, 2002, V. 50, 1639–1651.

71.   Sakai S., et al., *Acta mater.,* 1999, V. 47, No. 1, 211–217.

72.   Zhu Y.T., Huang J. Properties and nanostructures of materials processed by SPD techniques, in: Ultrafine Grained Materials II. (Eds Y.T. Zhu, T.G. Langdon, R.S. Mishra et al), Warrendale, TMS, 2002, 331–340.

73.   Amirkhanov N.M., et al., *Fiz. Met. Metalloved.,* 2001, V. 92, No. 5, 99–107.

74.   Koneva N.A.,et al., in: Physics and nanotechnology of functional nanocrystalline materials, V.1, (Ed. V. Ustinov and N.I. Noskova), Ekaterinburg, Ural Branch of Russian Academy of Sciences, 2005, 9–19.

75.   Noskova N.I., Mulyukov R.R., Submicrocrystalline and nanocrystalline metals and alloys, Ekaterinburg, Ural Branch of Russian Academy of Science.

76.   Koneva N.A., et al., in: Structure, phase transformations and properties of nanocrystalline alloys (Ed. Noskova N.I.). Ekaterinburg: IMP UB RAS, 1997, 125–140.

77.   Kozlov E.V., et al., *Ann. Chim. Fr.*, 1996,V. 21, 427–442.

78.   Kozlov E.V., et al., *Voprosy materialovedeniya*, 2008, No. 3 (54), 51–59.

79.   Kim H.S., et al., *Acta Mater.*, 2000, V. 48, No. 2, 493–504.

80.   Kim H.S., et al., *Mater. Sci. Eng.,* 2001, V. A316, 195–199.

81.   Conrad H., *Mater. Sci. Eng.*, 2003, V. A341, 216–228.

The structure and mechanical properties of nanocrystals

2.1. Introduction

It is well-known that the high-angle grain boundaries intensively inhibit dislocation slip at low and moderate temperatures. This assumption is especially valid for the Wboundaries greatly differing from the special boundaries. In the physics of strength, the effects associated with inhibition of the grain boundaries, are described at the yield point (σ_y) by the well-known Hall–Petch equation [1,2]:

$$\sigma_y = \sigma_0 + kd^{-1/2},\qquad(2.1)$$

where σ_0 is the deformation resistance in the absence of grain boundaries, d is the grain size, k is the Hall–Petch coefficient. This concept, named the grain boundary hardening [3,4], is widely used to increase the strength characteristics of metals, alloys and steels [5–7].

Almost all industrial materials are polycrystals.

The reliability of operation and processing of metal products is determined by their strength and ductility properties. To a large extent they depends on the grain size of the polycrystalline aggregate, its texture and the structure of the grain boundaries. The average grain size determines the mechanical properties by the Hall–Petch equation (2.1). Similar relationships exist for the flow stress at different degrees of deformation and tensile strength. Relation (2.1) is satisfied over a wide range of grain sizes down to nanosizes. There are cases where the exponent in equation (2.1) can differ from ½. In pure metals, the coefficient k for ultrafine-grained (UFG) polycrystals may begin to decrease with decreasing d, and for nanograins ($d \leq$ 25 nm) it becomes negative [8–10]. In the dispersion-strengthened alloys and intermetallic compounds the effect of reducing k can be suppressed [11].

The grain boundaries play an important role in plastic deformation and work hardening. They are sources of dislocations and sinks for them. Sliding of the dislocations is inhibited by them. There are cases when the grain boundaries (GB) let shifts to pass through. The formation of annealing twins and the development of the deformation substructure usually starts from the GBs. Finally, for some types of boundaries there may be sliding which increases with increasing temperature. The GB structure is defined by a number of important characteristics. The GBs can be equilibrium and non-equilibrium. The former are divided into specific and general type boundaries, the latter may contain segregations the excessive free volume in comparison with the equilibrium volume. On the contrary, instead of the free volume the GB may be contain a constrained volume. The structure, location and size of the free volume largely determine the properties of the GBs [12, 13].

The areas of the grain bodies immediately adjacent to the grain boundaries have a particular structure and properties. The characteristics of the border layers, along with the structure of the boundaries themselves, are extremely important for the formation of the mechanical properties of the polycrystalline aggregate. Description of the mechanical properties of a polycrystal requires the development of a composite model of the grain.

Increasing the strength properties by grain refining is an effective technique which can be used for all types of metallic materials [14,15]. Therefore, in the physics of strength this issue has received considerable attention.

2.2. Classification of polycrystals on the basis of the grain size

In recent decades, great progress has been made in producing materials with different grain sizes. In an effort to use the grain refinement [1, 2] to increase the yield point and tensile strength of metallic materials different technologies for obtaining fine grains of a polycrystalline aggregate have been developed [16,17]. This made it possible to manufacture metallic materials with virtually any grain size. In the analysis of their properties it is necessary to use the classification of polycrystalline aggregates on the bases of the grain size [9,18–20]. The fact is that a change of the grain size changes, firstly, the defective structure of the grains and of the border zone and, secondly, the structure of the grain boundaries, and, thirdly, the internal stresses, sources dislocation and deformation mechanisms. In view of the physical mechanisms of plastic deformation of the polycrystalline aggregates the relevant classification of the polycrystals on the bases of the grain size is given in Table 1.2.

The change in the average grain size d in the range of 1 nm to 1 cm changes the grain structure, the structure of the grain boundaries and the defective structure of the grain volume. Schematically this is shown in Figure 2.1. Increasing grain size changes several parameters. Firstly, the crystal structure is changed. From the atomic pile-ups and amorphous structure there is a transition to distorted imperfect crystals and then to relatively perfect microcrystals and, finally, to large grains with the ideal crystal lattice. Second, the grain structure changes. Small crystals can contain areas of free or constrained volume, stacking faults and microtwins. The grains in the following size range are dislocation-free. Larger grain may contain individual dislocations, first partial, then complete. A dislocation

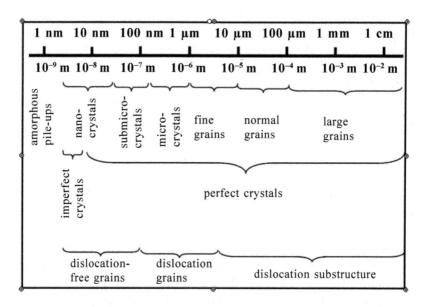

Fig. 2.1. Main special features of the structure of the polycrystal grains in the average grain size range 10 nm–1 cm.

substructure may form in sufficiently large grains. Third, the growth of the grain size changes the structure of their boundaries. They become perfect.

2.3. Methods for producing ultrafine-grained and nanograin polycrystalline materials

Increasing the strength properties by grain refining is an effective technique which can be used for all types of metallic materials [21]. In the last decade, different grain refining processes were developed by metal physicists. The most effective of these are various variants of severe plastic deformation (SPD): 1) equal channel angular pressing (ECAP) [16,17,21–23] and its various modifications, for example, equal channel angular extrusion (ECAE) [24], matched ECAP (ECAP–M) [25] when the metal is pushed through a channel having the shape of a circle, a combination of extrusion and ECAP [26], dissimilar angular pressing [16,27], equal channel angular drawing [21,28]; 2) high pressure torsion (HPT) [29,30]; 3) various methods of deep rolling deformation [29]; 4) combined methods (torsion plus rolling) [31,32], shear under high pressure [33], and others. In principle, it was possible to reduce the grain sizes by least 2–4 orders of magnitude compared to the previously used fine grains [34].

As an example it should be noted that the ECAP method used for pure copper increased the yield point σ_y by more than an order of magnitude. When using only one of the strengthening methods such result is a record. This achievement, associated with the increase of yield point, is due to grain refinement by moving to ultrafine-grained (UFG) and even nanograin materials. From Table 2.1, showing the classification according to the grain size of polycrystals, it is clear what the grain sizes have been achieved in recent years by various techniques. This section includes the UFG polycrystals, submicron- and nanopolycrystals.

In the wide range of the sizes of UFG and nanograins it was shown that the Hall–Petch relationship is valid [35]. There was a question about the dependences of parameters of this relationship on the grain size, and then on the mechanisms of deformation and work hardening, resulting in the implementation of the Hall–Petch relationship for UFG- and submicropolycrystals [8–10]. These issues are discussed below.

The current period of development of the science of the mechanical properties of the metallic polycrystals in the nanoregion is characterised by intensive development. This is well illustrated by numerous review papers and recent monographs [21,23,36–40]. Grain refining grains increases the yield point and the flow stress of metallic polycrystals making their use in industry very promising.

2.4. The structure of polycrystalline materials

The polycrystalline aggregate is a compact, solidly bonded set of single crystals. Within a single crystal the lattice retains its orientation except for small violations caused by the defective structure of the grains. It may be vacancies, interstitial atoms and their groups or pile-ups, stacking faults (SF), dislocations, dislocation pile-ups, low-angle dislocation walls. The polycrystal is an association of differently oriented single crystals of different sizes. Being combined in a polycrystal, separate single crystals, or crystallites, are referred to as polycrystalline grains. The polycrystal and its individual grains are three-dimensional spatial formations. Examples of polycrystalline aggregates of different materials and different grain sizes are shown in Figs. 2.2, 2.3 and 2.4.

A series of micrographs of polycrystalline aggregates is a flat section of a polycrystal. The pictures show the grains, separated by boundaries – these are the grain boundaries. Each grain has its own individual spatial orientation of the crystal lattice axes, which varies discretely when passing through the grain boundaries to the adjacent grain. The grains have different facets: there are polyhedral, spherical, ellipsoid of revolution facets and so on, including combined ones. The grains can have isotropic and anisotropic shape, elongated along

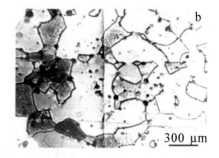

Fig. 2.2. Structure of the polycrystalline aggregate at the mesolevel: a) Cu+0.4 at.% Mn; b) Cu+40 at.% Zn.

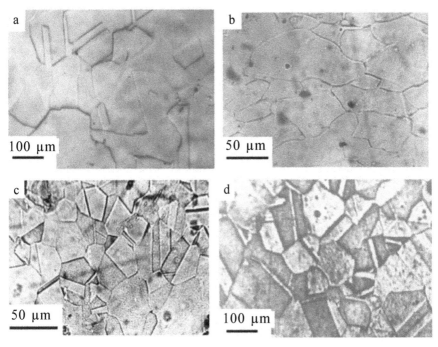

Fig. 2.3. The structure of polycrystalline aggregates at the mesolevel: a) copper; b) BCC 0.15C–Fe steel; c) FCC 0.18C–16Ni–10Mn–Fe steel; d) FCC 0.18C–15Ni–Fe steel.

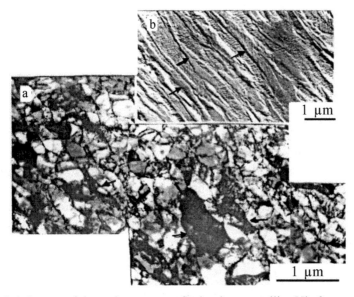

Fig. 2.4. Images of the grain structure of sub-microcrystalline Ni after equal channel angular pressing, produced by transmission electron microscopy (TEM) on the foils (a) and the replicas (b).

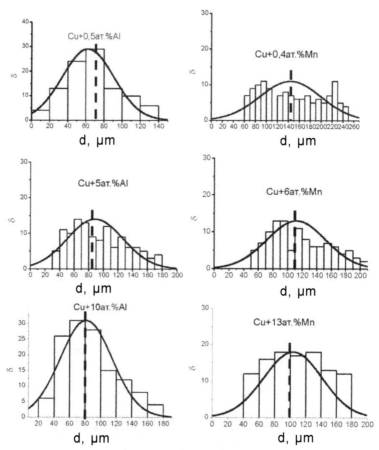

Fig. 2.5. Examples of the size distribution functions of the mesograins in the copper-based alloys.

one direction, for example, in the rolling direction. As can be seen, the GBs may be flat (in the section – straight), stepped (faceted) and curved. The grain sizes different from each other, so there are two features – the average grain size $<d>$ and the function of the grain size distribution $f(d)$ (see examples in Figure 2.5). The figure clearly shows that the distribution function can be monomodal, bimodal and polymodal. Figure 2.6 shows the function distribution for other grain size ranges.

As a result of technological treatments of different types the grain size of the polycrystalline aggregate can differ. In this case the volume of the sample group is characterised by the alternation of groups of grains with different local average size. The picture of such a polycrystalline sample is show in Fig. 2.7.

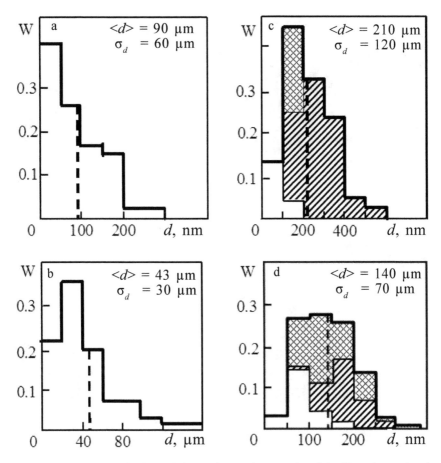

Fig. 2.6. The size distribution of the grains at the mesolevel (a, b) and nanolevel (c, d) in different materials: a) copper; b) BCC 0.5C–Fe steel; c) microcrystalline copper; d) microcrystalline nickel. In (c) and (d) the bright areas indicate the fraction of the grains not containing dislocation; – the fraction of the grains with the chaotic and network dislocation structure; – the fraction of the grains with the cellular dislocation structure.

The set of grains, or 'the ensemble of grains' of the polycrystalline aggregate is characterized on average by some parameters. They are the already mentioned average grain size $<d>$, the average anisotropy, or the anisotropy coefficient K (ratio $\ell/h = K$, where ℓ is the grain length, h is its width). The orientation of grains is described as the texture of the polycrystalline aggregate.

The set of the GBs of the polycrystal is characterized by its parameters. They are called the parameters of the ensemble of the GB.

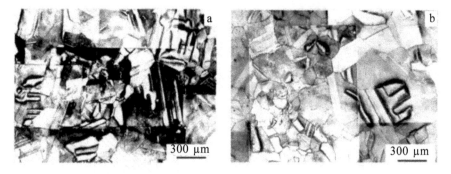

Fig. 2.7. Examples of polycrystalline aggregates with annealing twins and different grain sizes: a) Cu +3 at.% Al; b) Cu +5 at.% Al.

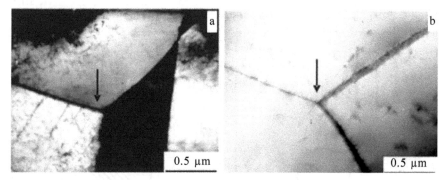

Fig. 2.8. Triple junctions of the grains in copper-based alloys: a) Cu +19 at.% Mn; b) Cu +5 at.% Al.

2.5. Triple junctions in grains

Three grains meet in the individual sections in the diagram in Figure 2.8. In the planar section these areas are referred to as triple junctions. In the actual three-dimensional polycrystalline aggregate there are areas in which the grains are joined along some spatial straight line (or curve) [41]. These lines are referred to as triple lines of the polycrystal (Figure 2.9a) [42]. In the three-dimensional polycrystal there are points where four grains meet. This object is referred to as the quaternary section (Figure 2.9b) or quadripole sections [43, 44]. In the triple and quaternary sections there may be partial or joint disclinations. In the flat patterns of the polycrystalline aggregates and thin foils the joint disclinations should be found in the triple junctions on the flat sections (Figure 2.10).

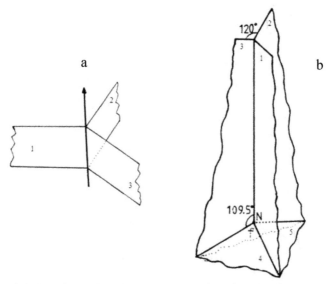

Fig. 2.9. Triple junctions (a) and quaternary section (b). The diagram according to V. Randle [42].

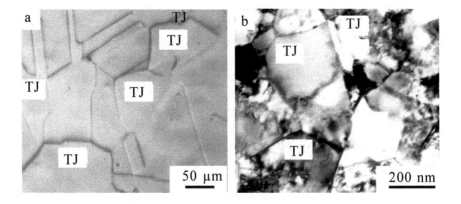

Fig. 2.10. Typical images of the triple junctions of the grains (TJ) in copper produced by optical (a) and electron (b) microscopy.

The quadripole sections are found seldom on the flat section of the surface of the specimen or in the thin foil, and the sections of the triple lines are found as triple junctions (Figure 2.8). The triple joint in this system for the polycrystals with the grains of the conventional dimensions ($d > 10$ μm) is classified on the basis of the types of grain boundaries (general type or special type) which are in equilibrium in them. In pure metals, there are often joints of the grain boundaries of the special type [41], and in the solid solutions there are quite often the joints of the grain boundaries of the general type which may include a single special boundary [45, 46].

The triple junctions play an important role in the nanograin structure because of several reasons [47]. Firstly, they are often connected with the partial disclinations [48–52] and are therefore sources of the stress fields and higher curvature–torsion of the crystal lattice. This fact was observed in experiments in [53, 54]. Secondly, the triple junctions influence plastic deformation, creep and fracture [55– 57]. Thirdly, the free volume, belonging to the grain boundaries [58, 59], in the triple junctions is concentrated to a larger degree than at the boundaries. Therefore, the triple lines are trajectories of faster mass transfer than the grain boundaries. Fourthly, the triple junctions are the areas of nucleation and formation of second phases [53, 60]. In the case of the non-equilibrium conditions of formation of these phases the triple junctions become sources of the field of internal stresses and greater curvature–torsion of the crystal lattice [53, 60, 61].

The nanopolycrystals are characterised by a small grain size ($d < 200$ nm), high density of the grain boundaries and triple junctions and, correspondingly, the complicated spectrum of the internal stress fields and the curvature–torsion of the crystal lattice.

The authors of [47] investigated the triple junctions of the grains in polycrystalline copper in the nanograin (grain size 130–320 nm) state. In these experiments, cylindrical specimens 0.2 mm long with a diameter of 10 mm were subjected to plastic deformation by torsion under hydrostatic pressure (THP) [62, 63] at a temperature of $T = 293$ K and pressure $P = 5$ GPa; the logarithmic strain was $\varepsilon = 5$. The chemical composition of the initial material was as follows, wt.%: Cu – 99.7; Ni – 0.10; O – 0.04; Sn – 0.05; Sb – 0.005; Fe – 0.05; P –0.02; Pb – 0.01; S – 0.01; As – 0.01; N – 0.005.

The experimental specimens were examined by transmission diffraction electron microscopy. The foils for electron microscopy (EM-125 and EM-125K microscopes) were produced by electrolytic polishing in the special conditions, producing large areas for examination. The thickness of the areas of the foil in examination in the electron microscope was ~170±25 nm, and was determined on the basis of the thickness extinction contours. The results, obtained from the sections of the specimens with the area of ~80 μm^2, containing 500–1000 grains, were statistically processed. Phase analysis was carried out by microdiffraction and the dark field method. The sizes of the grains, fragments, particles of the second phases and the scalar density of the dislocations were measured by the secant method. The data for $(1-3) \cdot 10^3$ triple junctions of the grains were statistically processed for each experimental point. The curvature–torsion (χ) of the crystal lattice was measured using the method based on measuring the parameter of the bending extension contours observed on the electron microscopic images of the structure [61, 64]. The value of χ was determined by the gradient of continuous disorientation:

$$\chi = \frac{d\varphi}{dl}, \tag{2.2}$$

where $d\varphi$ is the change the orientation of the reflective foil plane and dl is the displacement of the contour.

To determine χ it is necessary to measure the speed of movement of the extinction contour on the electron microscopic image at change of the angle of the goniometer or the width of the extinction contour. During special experiments it was established that the width of the contour in the disorientation values for copper and its alloys is ~1°.

In a deformed crystal the characteristics appearing in (2.2) are local. If there are no dislocations in the studied area, elastic bending-torsion is observed. Plastic bending-torsion is provided by the local excess dislocation density: $\rho_\pm = \rho_+ - \rho_-$ (ρ_+ and ρ_- are the densities of dislocations of different sign). The method used in the experiments allows one to define several components of the tensor of the elastic field, identify sources of long-range stresses, measure the amplitude of curvature–torsion of the lattice and establish patterns of change of the amplitude with increasing distance from the source of the stress.

Shear deformation in the local area of the sample was calculated from the formula

$$\varepsilon = 2\pi NR/l, \tag{2.3}$$

where N is the number of revolutions; R is distance from the centre of rotation; l is thickness of the sample.

During the production of samples by THP particles of secondary phase form on the defects of the structure. They are formed by both impurity elements in the initial copper and by capture elements from the environment (air). The free volume of lattice defects plays a large role in these processes. The elements forming the secondary phase arrive in two ways at the grain and fragment boundaries:

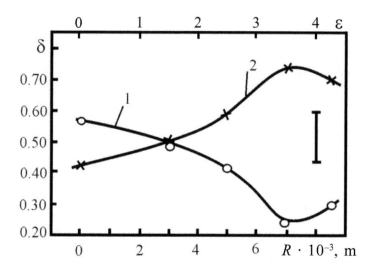

Fig. 2.11. Dependence of the size l (1) and volume fraction δ (2) of the Cu_3N particles in nanocrystalline copper on the distance R from the centre of torsion and on shear strain ε; 1′ and 2′ are the dispersions of the appropriate quantities.

transferred by sliding dislocations and diffuse in the free volume through defects.

Despite the large number of different types of nanoparticles formed in nanocrystalline copper in THP, triple junctions (usually at joints of fine dislocation-free grains) contain only Cu_3N phase particles. The size of these particles is 12–53 nm. The grain boundaries usually contain particles of Cu3 (Sn, Sb) and Cu_2O, and on the boundaries of the fragments and dislocations – Cu_2O particles. Note that Cu_3N is a non-diagrammatic phase, which is formed in the conditions of a large number defects due to the nitrogen present in the initial copper or trapped from air during SPD. From the centre to the edge of the sample the size of particles and their volume fraction decrease (Fig. 2.11).

It has been established that intensive interaction of the substitutional and interstitial impurities in metals with the crystal structure defects during SPD leads to both the formation and destruction of secondary phases of both the diagrammatic and non-equilibrium type [53, 60, 61].

In accordance with the obtained electron-microscopic images of the structure of nanocrystalline copper after THP the following classification of triple junctions was proposed: 1) joints with disclinations; 2) joints with stressed nanoparticles; 3) 'clean' joints

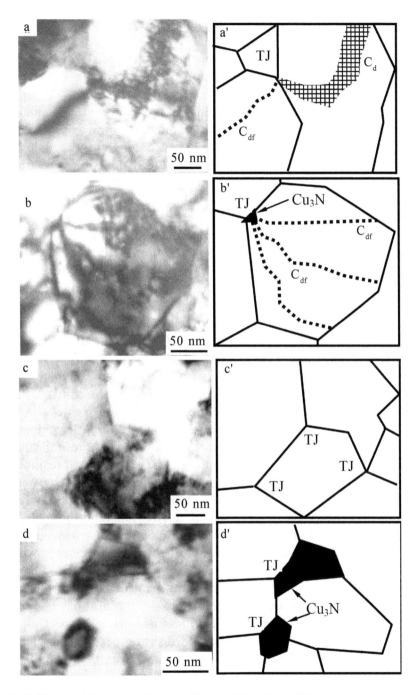

Fig. 2.12. Electron microscopic images of the triple junctions of the copper nanograins (a–d) and their appropriate diagrams (a'–d'): TJ – triple junction; C_d, C_{df} – dislocation and dislocation free contours.

(without particles and without disclinations); 4) joints with unstressed particles (Figure 2.12).

At the junction with disclinations there exists a discrepancy of the crystallographic orientation of the grains in contact. This discrepancy is compensated by a partial disclination. Typically, the electron microscopic images show extinction contours emanating from such junctions or surrounding them. A typical picture along with the scheme is shown in Figure 2.12a, a'. Here, the dislocation-free extinction circuit passes through the dislocation-free grain, in which there is no screening of the junction field. The dislocation contour passes through the grain with the dislocation structure, which partially shields the junction field.

At the junction with the nanoparticle, which creates an elastic field, there are extinction contours (Figure 2.12b, b').

In the *'clean' junction* there is no discrepancy, and when viewed in the electron microscope there are also no bending extinction contours. (Figure 2.12c, c').

The junction with disclinations and particles mutually shielding or compensating each other (Fig. 2.12d, d') has the discrepancy with a particle and there is no contour.

The junction with the unstressed particle can not be distinguished from the fourth type of junction by electron microscopy.

Junctions containing disclinations or stressed particles are one of the most important sources of stress. Only the 'clean' junctions with compensated disorientation do not create internal stresses. In electron-microscopic examination the stressed triple junctions in contrast to the unstressed ones are accompanied by bending extinction contours. In the studied nanocrystalline copper increasing strain decreases the proportion of unstressed junctions while the share of stressed junctions increases (Fig. 2.13). At a distance of $R = 7 \times 10^{-3}$ m from the centre of torsion of the sample the proportion of the stressed joints is 0.75, i.e. 2/3 of all junctions. It should be noted that recovery processes may take place near the edges of the sample

The main types of triple junctions in the studied copper are junctions with disclinations, the junctions with the particles and 'clean' junctions. The proportion of different types of triple junctions depends upon local deformation of the foil area being studied, i.e. on the distance from the centre of rotation of the samples. 'Clean' junctions are found mostly in the centre of the sample (Fig. 2.14). The number of junctions with disclinations and junctions with Cu_3N particles in the centre of the sample is somewhat less. With increasing

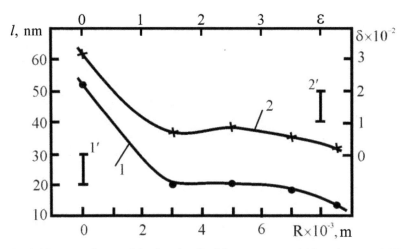

Fig. 2.13. Dependence of the fraction δ of the non-stressed (1) and stressed (2) triple junctions in nanocrystalline copper on the distance R from the centre of torsion of the specimen and shear strain ε.

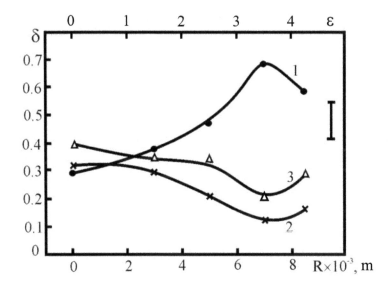

Fig. 2.14. Dependence of the fraction of the junctions δ of different type in nanocrystalline copper on the distance R from the centre of torsion under shear strain ε: 1) the junctions with the disclinations; 2) the junctions with the Cu_3N particle; 'clean' junctions (without the particles and disclinations).

degree of deformation, i.e. when moving from the centre to the edge of the sample, the proportion of junctions with disclinations rises sharply the proportion of junctions with particles and 'clean' joints decreases. When $R = 7$ mm the proportion of junctions with the disclinations is almost 3.5 times higher than that of the clean junctions and 7 times higher than the proportion of junctions with particles. At the edge of the sample due to the recovery and recrystallization processes the form of the dependence $d = f(R)$ is replaced by the reverse (see. Fig. 2.14).

Comparison of the data in Figs. 2.11 and 2.14 points to the evolution of the structure of nanocrystalline copper with increasing strain. This increases the density of disclinations in the junctions and decreases the proportion of defect-free joints, and the particles of the secondary phases are destroyed and dissolved. The phase-forming elements contained therein are moving to structural defects – dislocations, sub-boundaries, grain boundaries and the free volume of the triple junctions. The mechanisms of the screening of the field of internal stresses generated by triple junctions will be discussed.

The internal stress fields and curvature–torsion χ of the lattice are proportional to each other. The curvature–torsion amplitude is determined by its sources and screening mechanisms. The curvature-torsion of the crystal lattice was measured in the vicinity of triple junctions (at a distance of no more than 10–20 nm from the junctions). According to the information received, screening by the dislocation structure of internal stress sources significantly reduced the value of χ. At the same time, the point or line source, such as a triple junction, screens more efficiently than a flat source, such as the grain boundary (GB). So, as a result of screening the triple junction the value of χ decreased from 18.5×10^{-5} to 6.6×10^{-5} m^{-1}, while screening with the GB – from 9.4×10^{-5} to 4.9×10^{-5} ($\pm 3 \times 10^{-5}$, m^{-1}).

Analysis of the experimental data obtained by electron microscopy shows that elastic fields of the triple junctions may be screened by at least three methods [53, 61]:

- the dislocation structure inside nanograins that is possible only in fairly large grains [65–67];
- inside the triple junction with both the disclinations and the particles of the secondary phase (see Fig. 2.12); absence of the extinction contours on the electron microscopic image shows the mutual internal shielding of the particle and the disclinations in the junction (see Fig. 2.12d, d');

- symmetrical arrangement of disclinations of different signs at triple junctions.

According to the results the proportion of triple junctions, not screened by the dislocation structure is 5–6 times higher than of the screened junctions, and it increases to the edge of the sample (Fig. 2.15, the curves 1 and 2). This is a feature of the nanograin structure. When the grain size is larger than 10 μm, this ratio is rather reversed [68]. The small grain size, a large proportion of dislocation-free grains and underdevelopment of the dislocation substructure, which is typical of nanomaterials, complicate the shielding of the internal stress fields from the triple junctions by the dislocation structure.

Data on the number of the junctions with the stressed and unstressed particles are shown in Fig. 2.15 (curves 3 and 4). In the centre of the sample the proportions of the screened particles of Cu_3N, located at triple junctions, and unshielded identical particles. With an increase of deformation the proportion of the screened triple junctions with particles gradually decreases. This is due to the destruction by the moving dislocations of the stressed Cu_3N particles located in triple junctions. Comparison of the data in Figs. 2.14 and 2.15 confirms this: the type and screening of the triple junctions greatly change from the centre of the specimen to its edge. The amplitude of the curvature–torsion χ of the crystal lattice generated by different types of joints, changes only slightly (Fig. 2.16).

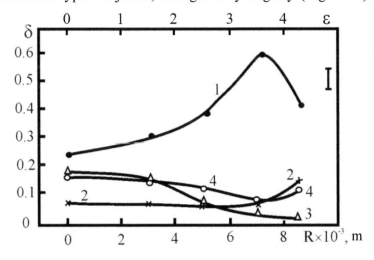

Fig. 2.15. Dependence of the number of the junctions δ of different type in nanocrystalline copper on the distance R from the centre of torsion and shear strain ε: 1, 2) the amount of the junctions not screened and screened by the dislocation; 3, 4) the fraction of the junctions with the non-stressed and stressed Cu_3N particles.

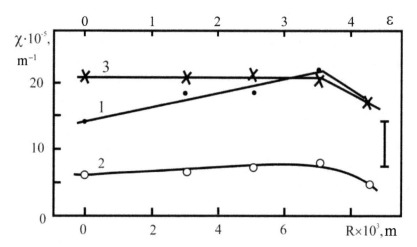

Fig. 2.16. Dependence of the curvature–torsion of the crystalline lattice χ in junctions of different type on the distance R from the centre of torsion and shear strain ε; nanocrystalline copper; 1, 2) – the junctions not screened and screened by the dislocations; 3) the junctions containing Cu_3N particles.

The dislocation structure shields the disclination junctions and reduces the amplitude of curvature–torsion of the crystal lattice approximately three times (see. Fig. 2.16). The values of χ for the case of unshielded disclination junctions and junctions with the Cu_3N particles are comparable.

2.6. Models of polycrystalline grains at the meso- and microlevel

Areas of bodies of the grains, directly adjacent to the grain boundaries, have a special structure and properties. The characteristics of the border layers together with the structure of the boundaries themselves are extremely important for the formation of the mechanical properties of the polycrystalline aggregate. Therefore, the description of the mechanical properties of a polycrystal at the mesolevel required the development of a composite model of the grain. The composite model of the grain for the period of its development absorbed all important aspects of a polycrystalline aggregate. At the same time, the analysis shows that in spite of the great advances of the composite model its development is not yet complete and the integrated model is not yet available. Besides it should be borne in mind that the Hall–Petch ratio was obtained by many authors using different models, including those not in the list of composite models [69, 70]. Therefore, the problem of the Hall–Petch relation in various grain models deserve further research attention. In some cases, the Hall–Petch ratio (2.1) has the form:

$$\sigma = \sigma_0 + kd^{-n} \qquad (2.4)$$

where n varies from 1 to ½.

The Kocks–Hirth model [71,72]. This is the most famous and at the same time the simplest model of a polycrystal in the group of the composite models. This model postulates that the grain is a composite (Fig. 2.17).

Grain size d is divided into two areas: $d = d_b + d_g$, where d_b is the hardened border zone with resistance to deformation τ_b different from the resistance τ_g of the internal part of the grain d_g. Using this hypothesis is possible to obtain the Hall–Petch type ratio (2.2). The mechanism of strengthening the border area is not discussed in this model.

The Arkharov–Westbrook model (for more detail see the monograph [73]). It is known that the due to the defective structure and the free volume the GBs absorb a significant amount of substitutional and interstitial impurities. This equilibrium segregation of impurities at low temperatures is concentrated directly at the boundary. At high temperatures, the elements are distributed evenly in a solid solution. An intermediate case occurs most often – increased concentration of the impurities in the grain and falling concentration profile with increasing the distance from the GB (Fig. 2.18.). Because thanks to the solid-solution strengthening the border zone is strengthened, then on both sides of the GB there forms intermediate (border) hardened zone (IBHZ in Fig. 2.18). The thickness of the GB is typically 0.5...2.0 nm [34, 74, 75]. The width of the IBHZ due to solid-solution strengthening by the impurities is usually much larger. According to [76], it can reach 40 nm. THe presence of the IBHZ is confirmed by numerous measurements of the microhardness on the GBs and its surroundings.

The Mughrabi model [77, 78]. If the GB has a specific dislocation structure, one can use the composite model proposed by H. Mughrabi. The dislocation density inside the grain is small, almost all the dislocations are concentrated in the border zone (Fig. 2.19). In

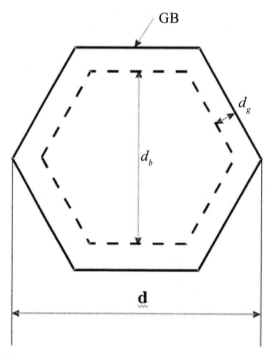

Fig. 2.17. The diagram of the grain in the Kocks–Hirth model: GB – grain boundary, d – grain size, d_b – the hardened grain boundary zone, d_g – the internal part of the grain.

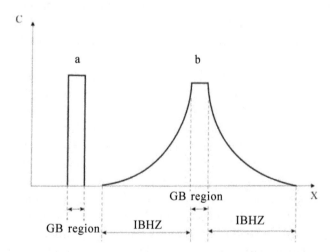

Fig. 2.18. The Arkharov–Westbrook model: a – the region of the grain boundaries; b – the grain boundary with the alloyed boundary regions; C is the concentration of the impurity; X is the distance.

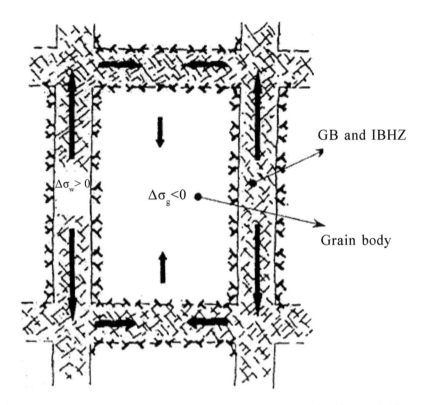

Fig. 2.19. The diagram of the grain (fragment, cell) in the Mughrabi model. The internal and surface dislocations can be seen. The arrows indicate the directions of the effect of the internal stresses ($\Delta\sigma_g$ – in the grain boundary, $\Delta\sigma_w$ – in the dislocation wall). The length of the arrows is proportional to the amplitude of internal stresses.

this model, the boundary and the boundary zone are not divided, but united in the wall with a high dislocation density. The internal stress fields in the wall $\Delta\sigma_w$ and in the grain $\Delta\sigma_g$ have the opposite sign, and the stresses in the wall are higher. The resistance to movement of dislocations through the grain body is much lower than through the wall, as in the wall the dislocation density and internal stresses are higher. For this model in the relation (2.2) $n = 1$. The variant of the Hall–Petch relation for the cells [79] r ratio for the cells [79] or the Hirth relation [72] operates at these boundaries..

  The model of strengthening by dislocations. For the first time Ashby drew attention to the fact that to preserve the integrity of the polycrystal it is necessary to increase the dislocation density in the vicinity of the GB [80, 81]. N.A. Koneva et al justified the higher dislocation density near the GB by the effect of accommodation systems in the vicinity of the GB [82]. The model of strengthening by dislocations was developed by Thomson and Moore [83,84]. Meyers et al [85] showed that the strengthening of the border regions by the dislocations is possible even in the elastic deformation stage of the specimen.

The model proposed by Koneva et al is based on the experimental data [86–89]. It is a continuation of the model constructed by Hirth. The model is suitable for both the yield point and for the developed plastic deformation for any grain sizes. The authors confirmed by direct experiments differences in the dislocation structure of the body of the grain and the boundary zone. The scalar density of the dislocations ρ in the boundary zone is higher and the size of the cells and fragments D is smaller than in the body of the grain [86]. Since the most powerful sources of internal stress fields (steps at the grain boundaries, grain junctions, which are disclination formations), are localised at the grain boundaries, the long-range stress fields τ_e, the curvature–torsion χ of the crystal lattice and the excess dislocation density $\rho_{\pm}$ ($\rho_{\pm} = \rho_+ - \rho_-$) are higher in the vicinity of the grain boundary than in the body of the grain. With increase of the distance from the grain boundary within the limits of the boundary zone the main parameters of the substructure (ρ, D, χ, $\rho_{\pm}$) gradually approach the average values in the body of the grain (Fig. 2.20). It may be seen that in this model the transition zone does not have a sharp boundary with the body of the grain, and the parameters

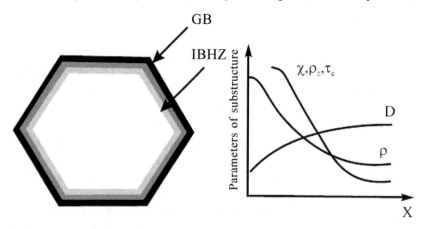

Fig. 2.20. Diagram of the grain of the polycrystal in the Koneva model: a – the grain; b – the behaviour of the parameters of the substructure in the boundary hardened zone, χ – the curvature–torsion of the crystal lattice; $\rho_{\pm}$ – excess dislocation density; τ_e – internal stresses; D – the cell size; ρ – the scalar dislocation density; X – the distance from the grain boundary.

of the substructure, responsible for the hardening of IBHZ, smoothly change to the average values of the body of the grain. The thickness of the IBHZ in the conventional grains was evaluated by the authors of [86, 87, 90] as equal to several microns. In this model, the Hall–Petch relation has the following form:

$$\sigma = \sigma_0 + k_1 d^{-1} + k_2 d_{-1/2} \qquad (2.5)$$

This relation efficiently describes the well-known fact [88] that in equation (2) the value of n changes from ½ to 1.

The authors of the model verified whether the main assumptions of the model are satisfied for the polycrystals at the microlevel. It was shown that the internal stress fields decrease from the boundaries of the grains and sub-grains in the same manner as in the materials with the conventional grain size, but the thickness of the IBHZ in transition to the microlevel greatly decreases [89]. According to the estimates of the authors, the size of the IBHZ is 20...30 nm. For the grains of the microlevel IBHZ may include the entire body of the grain. These fact indicates the general nature of a number of mechanisms ensuring fulfilment of the Hall–Petch relation in a wide range of the grain sizes [89–91].

As a result of the rapid development of the investigations of the polycrystals at the microlevel, composite models have been proposed for fine-grained polycrystals. They develop the concepts previously proposed for coarse-grained materials. The composite models are suitable for transition to examine in the microlevel. The polycrystals with this grain size are characterised by the higher defectiveness of the grain boundaries and their junctions [92] and the special type of distribution of the fields of the internal stresses in the vicinity of the grain boundaries [8, 20, 89]. It was also shown that the higher defectiveness of the grain boundaries of the microlevel causes that they are subjected to plastic deformation at room temperatures by the grain boundary sliding with the increase of the intensity of the diffusion processes in their vicinity [8, 93, 94].

The 'mantle' model. The structure of the grain boundaries and the internal stress fields and also segregations lead to the formations of the hardened zone in the vicinity of the grain boundaries. In the conventional polycrystals at the mesolevel, this hardened zone occupies a small fraction of the grain body. This fraction increases with a decrease of the average grain size and at the microlevel may reach half the grain size, and in the region of the nanocrystals

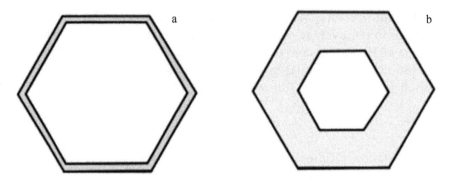

Fig. 2.21. The relative size of the hardened grain boundary zone forming the hardened mantle around the grain body: a) mesopolycrystal; b) micropolycrystal.

it can include the whole grain. This is schematically described quite efficiently by the model of the grain boundary mantle [95]. The diagram of the variation of the relative size of the hardened zone, including the grain boundary, is shown in Fig. 2.21. In fact, this scheme repeats the Kocks–Hirth model described in [9], for nanopolycrystals.

The model by R.Z. Valiev et al [17]. In developing the model, the authors used their results of X-ray diffraction analysis and electron microscopic and information on the elastic fields, calculated for different dislocation– disclination configurations in polycrystals at the microlevel. In this model, special emphasis is placed on the following special features of the structure of the grain boundaries and the boundary zone. Firstly, it is the defective structure of the non-equilibrium grain boundaries, containing dislocations (Fig. 2.22a) [17, 84]. These are the lattice dislocations. Some of them can glide in the plane of the grain boundaries, others are sessile dislocations. Secondly, disclinations of different power are located in the grains junctions (Fig. 2.22). The elastic fields from the dislocations and disclinations extend into the boundary region. These distortions gradually attenuated with transition to the body of the grain [17, 97]. According to the estimates by the authors, the thickness of the IBHZ is 10...20 nm.

Three-dimensional composite models. The investigation of the structure of the grains at the microlevel requires the development of new more complicated models of the composite structure of the grains. If planar two-dimensional grain models were sufficient for the polycrystals at the mesolevel, three-dimensional composite models were proposed for the UFG and nanopolycrystals at the

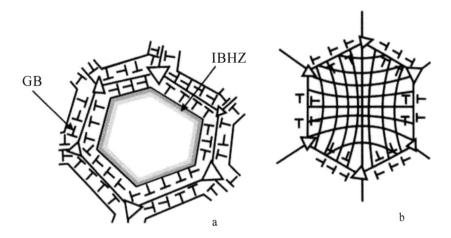

Fig. 2.22. The grain of the ultrafine-grained material in the model constructed by R.Z. Valiev et al with the size of approximately 100 nm (a) and 10–20 nm (b); the triangles of different size indicate the disclinations of different power; the distorted lines in (b) show the distortions of the crystal lattice. To provide more information, the thickness of the grain boundary is slightly exaggerated.

microlevel. The studies of this problem are still continuing, starting with the work by Palumbo et al [98]. In this end subsequent studies, Suryanarayana et al [99], Wang et al [100] developed a four-components composite model. The model includes in detail the grain with all adjacent boundary formations. This model contains the grain boundary on which sliding can take place, as in the creep of conventional polycrystals. The model also takes into account the presence of triple junctions and quadripole sections. Although there is no IBHZ in this model, the composite nature of the model is ensured by definition of four areas of the material with different mechanical characteristics at a sufficiently large width of the grain boundary, triple junctions and quadripole sections. The second special feature of the model is that in addition to hardening, it also considers softening. Softening is caused by the low activation energy of self-diffusion on the grain boundaries of the polycrystals at the microlevel [8]. The theory of plastic deformation and deformation resistance for this model was developed by Kim, Estrin and Bush [43, 44].

The Kim–Estrin–Bush model. To describe plastic deformation, the authors of [43, 44] into account the dislocation sliding and both lattice and grain boundary diffusion. In this model and its computer variant for Cu it is possible to describe the conventional behaviour according to Hall–Petch in accordance with the equation

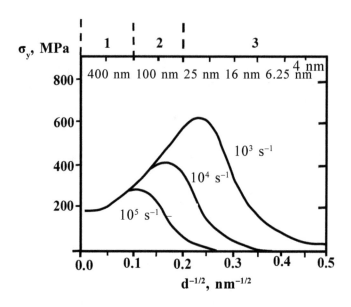

Fig. 2.23. Dependence of the yield point on $d^{-1/2}$ at different strain rates: 1) the region of dislocation sliding; 2) lattice diffusion; 3) grain boundary diffusion.

(2.1) for the grains larger than 50 nm. In the grain size range 50...16 minute nanometre, the parameter k decreases and then changes to the negative values (Fig. 2.23]. In modelling the mechanical properties the model takes into account different volumes and parameters of the body of the grain, the grain boundaries 1 nm wide, triple junctions and quadripole sections. The decreasing grain size increases the volume fraction of the defective material, porosity and thermally activated processes become easier. At the same time, dislocation sliding as the main mechanism of deformation is replaced by a grain boundary diffusion. The role of volume diffusion mass transfer in the transition region between them increases. All these results with grain refining in the decrease of the parameter k and a subsequent change of its sign. Although there is no IBHZ in this model, the composite nature of the model is ensured by the definition of four areas of the material (Fig. 2.24) with different mechanical characteristics at a sufficient width of the grain boundary, the triple junctions and quadripole junctions. The contribution of dislocation sliding along the grain boundaries is not taken into account in this model. In [8, 93] calculations were carried out to determine the quantitative parameters of intragranular and grain boundary sliding for certain

Volume of
boundary

Grain body

Triple junction

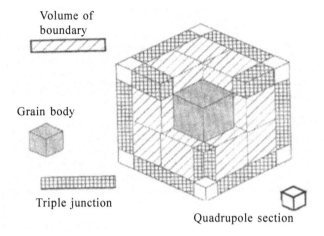

Quadrupole section

Fig. 2.24. The three-dimensional composite model of the grain according to Kim, Estrin and Bush[43, 44].

materials with a grain size at the microlevel. The experiments show that an especially large contribution of the grain boundary sliding is organised on the selective game boundaries delineating the entire groups of the grains. In most cases, these boundaries are oriented in the vicinity of the maximum acting cleavage stresses.

2.7. The structure of individual nanograins

In the standard mechanisms of formation of the grain, the grain consists of the grain body and the boundary regions. The average grain size has a strong effect on the crystalline structure of the body of the grain and its boundaries (Table 2.1). The structure of the nanograins is a nanocrystalline formation characterised by the equilibrium crystal lattice. The perfection of the lattice is impaired by decreasing grain size. At a grain size 5 nm and less the lattice is greatly distorted. At the grain sizes smaller than 3 nm, the non-equilibrium or quasi-equilibrium crystal structure can form.

Another factor distorting the crystal lattice of the grain is the dislocation structure. The formation of the misoriented boundaries of the cells and low-angle boundaries of fragments (blocks or sub-grains) results in the appearance of misorientation on one part of the grain in relation to the other one, up to several degrees.

Finally, the third reason for the distortion of the crystal structure of the grain body is the presence of internal stresses. They form

Table 2.1. The structure and internal stress fields of amorphous and crystalline materials in a wide scale range

No.	Classification	Range of scales or grain size	Defective structure after preparation of material
1	Pile-ups, amorphous state	0.5–2 nm	Amorphous structure. Mean coordination number differs from crystalline. Concentrational delamination
2	Imperfect crystals	2–3 nm	Highly imperfect orders, approaching crystalline. Atomic spacing differs from crystalline and is characterised by considerable dispersion. The quasilattice is greatly distorted. Concentrational delamination may occur. Strong dependence of physical properties on the crystal size. Start of formation of the granular structure
3	Nanocrystals	3–10 nm	Crystal lattice almost perfect. Crystal boundaries blurred. Lattice parameter ceases to depend on crystal size
		10–50 nm	Partial dislocations, microtwins
4	Submicrocrystals	10–100 nm	Almost defect-free crystals, separated by highly defective boundaries with point defects, their pile-ups and dislocations. Disclinations at grain junctions
		50–100 nm	Total dislocations
5	Microcrystals	0.1–1 μm	Crystals may contain dislocations and also substructure (cells, fragments). Different grain boundaries: imperfect defective, close to perfect of general type and perfect special.
6	Mesograin polycrystal	1–10 μm	Relatively perfect grains. Annealing twins. Substructure. Boundaries contain a few defects, mostly of general type, less frequently special
7	Conventional grains	10–100 μm	Fraction of special GBs comparable with that of general GBs. Ledges at special boundaries. Many annealing twins at low stacking fault energy. Low dislocation density
8	Large-grained polycrystal	0.1–1 mm and more	Small number of defects in initial condition in material

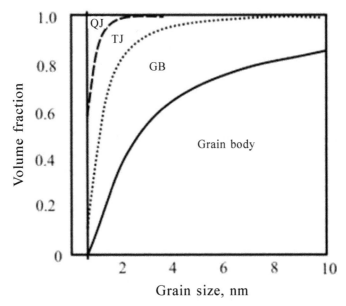

Fig. 2.25. Dependences of the volume fractions of different components of the structure of the grains on the grain size in the three-dimensional model. Notations: GB – the grain boundary, TJ – the triple joints, QJ – quadripole joints.

because of several reasons: 1. The non-uniform deformation of the grains of the polycrystalline aggregate; 2. The presence of second phase particles deformed by a mechanism which differs from that of the body of the grain; 3. The presence of dislocations in the body of the grains and at the grain boundaries; 4. The presence of partial disclinations, distributed in the grain junctions; 5. Ledges at the special boundaries; 6. The constricted volume of the grain boundaries. The internal stresses lead to the formation of a continuous misorientation of the crystal lattice of the grains.

The boundary region of the nanograins forms from three different types of the material of the polycrystalline aggregate characterised by different structures: 1) the flat grain boundary interlayers of a small thickness, representing the main volume of the grain boundaries; 2) the columnar structure of the triple junctions; 3) the quadripole sections, characterised by the isotropic structure. This terminology and these forms of the grain boundary formations were obtained as a result of describing the cubic grains.

The volume fractions of different areas of the body and the grain boundary in the polycrystalline aggregate depend, firstly, on the average grain size and, secondly, on the average thickness of the

grain boundary formations. At the mesoscopic grain sizes a large part
of the volume of the polycrystalline aggregate is occupied by the
grain body. The role of the grain boundary material increases with
a decrease of the grain size. With a further refining of the grains
the role of the triple junctions increases and then the role of the
quadripole sections (Fig. 2.25) [98–100]. It should be remembered
that the thickness of the grain boundary formations can change: 1]
with the change of the grain size; 2) the type of the grain boundary;
3) its proximity to equilibrium; 4) the presence of joint disclinations,
and 5) the component composition of the nanomaterial. At present,
both the experimental data and theoretical estimates and also the
simulation results indicate that the thickness of the grain boundary
areas is in the range 0.5...1.0 nm. However, in certain cases it can
be larger.

2.8. Special features of the structure of the nanopolycrystalline aggregate as a consequence of high plastic strains

The material after high plastic strains is a polycrystalline aggregate
consisting of the ensemble of grains, separated by boundaries. In
a general case, the grains have different sizes and orientation of
the crystal lattice and are surrounded by different types of grain
boundaries. According to their dimensions, the grains can be isotropic
and anisotropic. Sometimes, the grain groups are separated from
other groups by boundaries of the same, spreading along several
grains. Typical examples of sub-microcrystalline aggregates of nickel
grains, produced by the methods of equal channel angular pressing
and torsion under high quasi-hydrostatic pressure (see chapter 5)
are presented in Figs. 2.4 and 2.26. Electron microscopic images,
produced using the Ni foils, demonstrate the polycrystalline aggregate
with grains of different sizes and orientation. The orientation is
indicated by different contrast of the grains.

The typical size distribution function of the grains is shown in
Fig. 2.6. There are three types of grains and sub-grains usually
found after high plastic strains [10]. They are characterised by
different dislocation structures, namely: 1) dislocation-free grains;
2) the grains with the chaotic dislocation structure; 3) the grains
containing the dislocation substructure – cells or fragments. At
the average grain size of 140 nm the average size of each type of
grain increases in the direction from the dislocation-free grains to
the grains with the cells and fragments (Table 2.2). This structure

Table 2.2. The size of grains of different type in Ni produced by severe plastic deformation

Type of grains	Average grain size, nm	Dispersion, nm
Dislocation-free grains	84	84
Grains with chaotic dislocation structure	150	53
Grains with cells and fragments	166	57
All grains	140	60

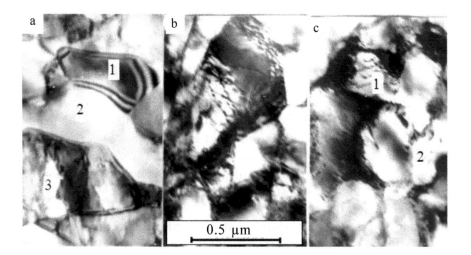

Fig. 2.26. Electron microscopic images of three types of grains in ultrafine-grained nickel after torsion under hydrostatic pressure: 1, 2) the grains without dislocations; 3) the grain containing fragments (a); groups of grains containing dislocations (b); 1 – the grains containing dislocations, 2 – the grains without dislocations (c).

of the grains and the size distribution of the grains are determined by the methods of preparation of the nanostructured polycrystalline aggregate. At some moment of time deformation was interrupted and, therefore, the structure contained dislocation-free fine grains and the grains of medium sizes with a developing dislocation structure. The largest grains of the nanocrystalline aggregate contained cells which transformed gradually to fragments during deformation, i.e., the subgrains surrounded by the low-angle boundaries. With further deformation the subgrains transform to new nanograins. Thus, the size distribution function of the nanograins illustrates the mechanism of their formation during deformation. The dislocation-free grains grow as a result of migration of the grain boundaries, and intragranular deformation leads to the formation in them of initially the chaotic dislocation structure and then the cellular structure. The cell boundaries become misoriented, and undergo transformation to low-angle sub-boundaries of the fragments. With further deformation the misorientation of the boundaries increases. They transform to the boundaries of the nanograins.

2.9. Dependence of the dislocation density on the grain size and the problem of fine grains without dislocations

The mechanical properties and the behaviour of the Hall –Petch coefficient at the microlevel are influenced by the size distribution of the grains, the internal substructure of the grains and the field of internal stresses. The substructure depends on the grain size. The largest grains contain dislocation cells, the grains with the mean size the chaotic dislocation structure. The fine grains are free from dislocations. The formation of the dislocation substructure depends on both the grain size and the internal stresses.

The interaction of the dislocations with the sources of internal stress fields and, in particular, with the grain boundaries has a strong effect on the dislocation structure and deformation mechanisms of the UFG materials. In the polycrystals at the microlevel the grain boundaries are not only sources of dislocations but also sinks for dislocations. The latter effect influences the dislocation density and leads to the formation of dislocation-free grains.

Figure 2.27a, b shows the dependence of the average scalar density of the dislocations (ρ), typical of the given grain size, on the average size (d) of the same type of grains. The data for Cu, Ni and the Cu–Al–O alloy are presented [8]. The increase of the values of ρ and d is accompanied by the transition from small almost dislocation-free grains to the grains with the chaotic substructure and then to the grains with the cellular substructure. The data for copper and the copper alloy for the freshly prepared (after severe plastic deformation) and the annealed condition (severe plastic deformation + annealing) fit a single curve. The appropriate data for Ni are separated. It is important to stress that the ultrafine grained and nanocrystalline states for the presented polycrystals at the microlevel were produced by different methods. Copper was prepared by torsion under hydrostatic pressure, nickel by equal channel angular pressing, the copper alloy by high-temperature drawing through dies. In all cases the dependences shown in Fig. 2.27 were identical. In the

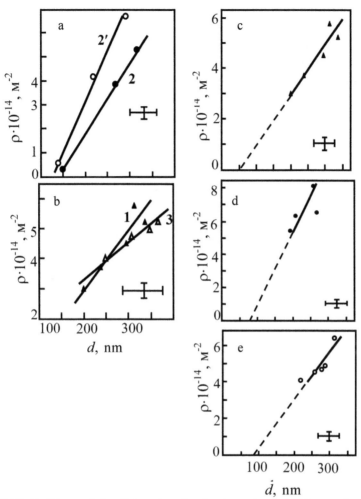

Fig. 2.27. Build up of the dislocation density in grains of different size (d) and type in ultrafine-grained polycrystals: a, b) dependence of the dislocation density (ρ) on the average grain size (1 – Cu; 2, 2' – Ni; in the case of Ni the straight line 2 corresponds to the freshly prepared state after ECAP; 2' – the condition after annealing; 3 – the Cu–Al–O copper alloy); c, d, e) dependence of ρ on d for the ultrafine-grained copper (c – for the grains with the average size, d – for the grains with the chaotic dislocation structure, e – for the grains with a fragmented substructure) in annealing; 3 – Cu–Al–O copper alloy); c, d, e – dependence of ρ on d for ultrafine-grained copper (c – for the average grain size, d – for the grains with the chaotic dislocation structure, e – for the grains with the fragmented substructure)..

range $d = 100...500$ nm the mean dislocation density decreases almost linearly with decreasing grain size. This effect is caused in particular by the intensive interaction of the dislocations with the grain boundaries and by the pulling of the dislocations into them.

The dependence $\rho = f(d)$ also indicates that the dislocation density ρ tends to the rolling in the vicinity of the grain size $d = 100$ nm. In Fig. 2.27 a similar dependence obtained by the authors of this book is presented for a number of materials (Fig. 2.27). The formation of the dislocation-free grains in the vicinity $d = 100$ nm results in reaching the second critical grain size (the first critical grain size was discussed in section 2.14).

A similar tendency is also indicated by the partial histograms of the size distribution of the dislocation-free grains. They are shown for copper, nickel and the copper alloy in Fig. 2.28. As indicated by the figure, the range of the size of the dislocation-free subgrains extends from the smallest size to $d \approx 200$ nm, irrespective of the type of material [8]. This indicates the same nature of the phenomenon. In the case of the copper alloy, the maximum of the distribution function is displaced to larger grain sizes. Here the difference in the preparation procedure of the materials and the solid solution hardening factor exert effects.

The dislocation-free grains play a special role in the formation of the mechanical properties of the sub-microcrystals [91]. Because of the small size, these grains hardened the polycrystal. The number

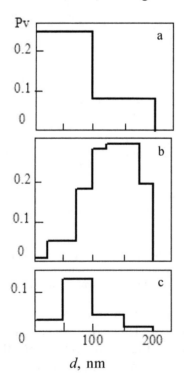

Fig. 2.28. Size distribution of the dislocation-free grains in the submicro-crystals of Cu (a) and the Cu-based alloy (b) and in submicrocrystalline nickel (c).

of the dislocation-free grains in relation to the total number of the grains varies in the range 0.10 to 0.14. The volume fraction of these grains does not exceed usually 0.05...0.20 because of the small dimensions.

2.10. Critical size ranges of the grains and areas with grains

The dislocation-free grains are observed when their size is smaller than the second critical size which for the pure metals is close to $d_{cr} \leq 100$ nm [10]. As a result of grain refining produced by severe plastic deformation, large changes take place in the structure of the polycrystalline aggregate. Firstly the density of the grain boundaries increases. Secondly, the high density of defects at the grain boundaries is retained. These are the defects of the dislocation and disclination type and also grain boundary steps. Thirdly, the defective structure of the body of the grains changes. The scalar dislocation density inside the grains decreases (Fig. 2.27]. Figure 2.27 shows that $d_{cr} \approx 100$ nm for pure metals. Fourthly, the fraction of the geometrically necessary dislocations in the dislocation structure increases. The internal stresses increase correspondingly. Fifthly, when reaching some critical grain size the dislocations leave the body of the grain and concentrate at the grain boundaries. Sixthly, the fraction of the ternary junctions of the grains increases and the dislocation density in them also increases. Seventhly, the change of the grain size increases the density of disclinations and their power and also the curvature – torsion of the crystal lattice. When the parameters of the critical grain structure are reached, the main type of defects – dislocations – changes to partial disclinations. When the grain size approaches the critical value, the intragranular dislocation density initially decreases and the grains then become dislocation-free. This is accompanied by an increase of the density of the partial disclinations at the grain boundaries and, in particular, at the triple junctions. At the average grain size of $d = 100$ nm, the dislocation structure in the nanopolycrystals is completely replaced by the disclination structure. The density of the disclinations reaches the values equal to the dislocation density in the deformed materials. At the grain size of $d \leq 100$ nm, the dislocation sliding in the nanograins steel takes place about the buildup of dislocation no longer curse. With a further decrease of the grain size the total dislocations are replaced by partial dislocations, twins, stacking faults, semi-symmetric sections of the free and constricted volumes.

The investigation of the properties of the nanopolycrystals showed that the different grain sizes, the sizes of their sections in grain boundaries ensure different mechanisms of the deformation processes, phase transitions, achievement of the equilibrium properties and can be critical. The main critical ranges of the grain size and of their areas are presented in Table 2.3 [101, 102].

2.11. The Hall–Petch relation and its parameter σ_0 in a wide grain size range

The problem of deformation resistance (σ) of polycrystals at the microlevel, produced by severe plastic deformation, is very important. Basically, the solution of this problem is concentrated in the analysis of the applicability of the Hall–Petch (H–P) relation, the behaviour of parameter k in a wide grain size range, and the examination of the experimental data and the deformation mechanisms. A number of models for analysis of the behaviour of parameter k has been proposed. The application of the effect of grain refining for increasing the deformation resistance and work hardening of the metallic materials is closely linked with the parameter k in the H–P relation. The problem of constancy or change of the parameter k and σ_0 in different grain size ranges is still in the centre of attention of investigators.

Table 2.4 gives the values of the parameter σ_0 for copper in a wide grain size range, from microns to nanometres.

The data were obtained in tests at room temperature. As expected, parameter σ_0 which characterises the deformation resistance in the body of the grains varies in a wide range. For larger grains, the parameter varies from 10 to 90 MPa, for the size range including nanometric grains is varies from 10 to 150 MPa. The variation of σ_0 for pure copper is determined by the presence of impurities in the crystal lattice and structural defects, mostly dislocations. Attention should be given to the fact that the ranges of variation of σ_0 for different grain sizes coinciding most cases. This indicates the same nature of the phenomenon.

2.12. The mechanisms of implementation of the Hall–Petch relation at the mesolevel

Prior to analysing the H–P relation at the microlevel, it is convenient to examine the nature of formation of this relation at the mesolevel.

Table 2.3. Critical sizes of the grains and their areas

No.	Physical factors and mechanisms	Grain size of width of GBs, nm
1	Formation of grains and their boundaries	2
2	Thickness of grain boundary	0.5–1.0
3	Width of the interlayer when grain boundary sliding and grain boundary diffusion take place	2–3
4	High-intensity deformation processes at the GBs and triple junctions	1–10
5	High-intensity deformation processes at the imperfect crystal lattice, displacement of free or constricted volume	2–10
6	Effect on diffusion phase transitions and martensitic transformations	2–5
7	Formation of equilibrium crystal lattice	3–5
8	Equilibrium volume properties	5–10
9	Nanograins	2–100
10	Transition to the negative Hall–Petch parameter. The first critical grain size	10–50
11	Special mechanisms of generation of dislocations by grain boundaries	10
12	Sliding of partial dislocations, twinning	10–100
13	Stable emission of dislocations by steps at GBs	≥ 25
14	Sliding of total dislocations in the grain	25
15	Submicrocrystals	100–500
16	Formation of dislocation pile-ups	50
17	Dislocation-free grains, pulling of dislocations to GBs. Second critical grain size	100
18	Sliding of perfect dislocations	>100
19	Formation of substructures	200
20	UFG grains	500–1000
21	Mesograins	>1000

This is especially important taking into account the available information according to which the calculated value of the parameter k [9], equal to 0.11 MPa $\cdot$ m$^{1/2}$, in equation (2.1) for copper remains unchanged when the average grain size varies from $2.5 \cdot 10^5$ nm to $4 \cdot 10^1$ nm. In a general case, the value of constant k from equation (2.1) depends on the type of material (metal or alloy, ordered alloy

Table 2.4. The Hall–Petch parameter σ_0 for copper in a wide grain size range

Grain size range, μm	σ_0, MPa	Reference
3...29 μm	60	[103]
2.5...20 μm	10	Data by the authors
10...60 μm	25	[104]
12...178 μm	90	[105]
20...250 μm	23	[106]
-	4.0	[107]
12...250 μm	95	[108]
3...150 μm	10	[109]
10...100 μm	11.4	[110]
17...1000 μm	13	[111]
1.4...11.6 μm	20	[112]
200 nm...50 μm	10	Data by the authors
11 nm...100 μm	11	[113]
300 nm...100 μm	30	[110]
25 nm...2.5 μm	100	[114]
200 nm...90 μm	150	[114]
16 nm...10 μm	75	[115]
16 nm...10 μm	130	[116]
25 nm...20 μm	80	[117]
26 nm...20 μm	70	[118]
35 nm...40 μm	25	[119]

or disordered alloys, the type of crystal lattice), its purity, test temperature, defectiveness of the structure and the average grain size. It was believed for a long time that the coefficient k in equation (2.1) is independent of the grain size [69]. Tables of values of k for different materials were also published (see the review in [35]). The ranges of variation of the parameter k, obtained by experiments, for the main metals are presented in Table 2.5. In this chapter, the dependence of parameter k on only one factor – the average grain size – is analysed.

Equation (2.1) can be derived for the composite model of the grain. Equation (2.1) was also derived for other models: 1) the models linking the role of the grain size with the stress concentration in the individual slip bands [1, 2, 120, 121]; 2) work hardening models based on the dependence of the dislocation density or path length of the dislocations on the grain size [122–125]; 3) the models based on the assumption of the controlling role of the surface and grain boundary sources of dislocations in the process of transfer of slip from grain to grain [122, 126]; 4) the models taking into account

Table 2.5. Ranges of variation of the Hall–Petch coefficient for different metals

Metal	k, MPa m$^{1/2}$
Al	0.02÷0.29
Cu	0.01÷0.24
Ni	0.12÷0.28

different orientations of the grains and the resultant either elastic anisotropy of the contacting grains [69, 127] or plastic deformation [62, 127] and, as a result, the incompatibility of their deformation; 5) the models taking into account the blocking of the dislocation cells in the grain by the reverse stress fields from dislocation pile-ups in the same grain [86–88] (this model is supplemented by the shear resistance of the scalar dislocation density of other systems, present in the grain); 6) the models based on the assumption according to which the shear in the grain in the vicinity of the grain boundaries accompanied by rotating deformation [129]. The models 5 and 6 gives the relation for the equation similar to the H–P relation and can be used to explain the presence in equation (2.2) of the constant n with the value between ½ and 1.

At present, the experimental data are not sufficient for complete analysis of all the models. However, certain generalisation can be made. The model 1 is realised in experiments far less frequently than the models 2–5. It is now well recognised that in the vicinity of the yield point the dislocations are emitted from the grain boundaries (model 3), the dislocation density is inversely proportional to the grain size (model 2), the main and accommodation sliding split dislocations into statistically stored and geometrically necessary dislocations, and the activity of the main and accommodation systems depends on the orientation of each specific grain and contact stresses inside the polycrystal (model 4). In other words, the experiments indicate that the models 2–5 are implemented together.

The models denoted above by the numbers 1, 2, 3 and 5 are based on the pile-ups concept. This concept is productive and can be easily verified by experiments. The experimental data and theoretical considerations indicate that the boundary regions at moderate and low-temperatures in the mesograins are hardened to a greater extent than the internal part of the grain body because the complex model is added to the above models. The geometrically necessary dislocations are concentrated especially in the boundary region. This area is characterised by the effect of the accommodation slip systems and there are dislocations inhibited by the grain boundaries. In fact, the compositional model, derived from the specific mechanisms, takes into account their presence in a generalised manner.

In addition to these specific models, the literature describes the systematic development of the composite model of the grain image the hardening of the body of the grain and the boundary region is evaluated differently. The diagram of the cross-section of the grain in the composite model is shown in Fig. 2.17. The experimental data and theoretical considerations indicate that the boundary regions at moderate and low temperatures in the mesograins are hardened to a greater extent than the internal region of the grain body. Initial studies of these models are associated with the names of Arkharov, Westbrook, Kocks, Hirth, Margolin and others. The boundary regions of the grains with the width d_g can be hardened as a result of the higher concentration of interstitial and substitutional impurities, as a result of the higher dislocation density and the restricted path length of the dislocations, as a result of the higher internal stresses in the vicinity of the grain boundaries (and their triple and quaternary junctions). A detailed spatial scheme of the grain is shown in Fig. 2.24. The volume of the grain body is hardened to a lesser extent than the boundary region. In fact, the Ashby model belongs to the group of the composite models.

The composite models have proved to be suitable for transition to examining UFG and sub-microcrystals. The polycrystals with this grain size are characterised by higher defectiveness of the grain boundaries, their junctions and the special type of distribution of the fields of internal stresses in the vicinity of the grain boundaries. At the same time, the results show that the higher defectiveness of the grain boundaries of the UFG materials and of sub-microcrystals causes that they are included in plastic deformation at room temperature by grain boundary sliding and diffusion processes in their vicinity [8].

2.13. Dependence of coefficient k on the grain size in the Hall–Petch relation

The problem listed in the heading of this section is a basic problem in the analysis of the dependence of the flow stress on the grain size in transition from the polycrystals at the mesolevel to the polycrystals at the microlevel. Although there is a number of theoretical concepts, describing this transition, the experimental data are controlling factor for this analysis. Initially, all the data were obtained by hardness measurements. This was followed by taking individual measurements on pure metals at the yield limit. At present, the data for polycrystalline copper, nickel, titanium, iron and aluminium are available for a wide grain size range. Measurements are taken on these metals include both the mesolevel and the microlevel.

The most widely used concept is based on the inflection point of the H–P relation $\sigma = f(d^{-\frac{1}{2}})$ on reaching the critical grain size d_{cr} (Fig. 2.29]. This is the first critical grain size. At $<d>$ greater than d_{cr}, the H–P constant $k > 0$. After reaching d_{cr} when $<d>$ is smaller than d_{cr}, the constant k becomes negative. This means that with further grain refining the yield point will decrease. The value d_{cr} was measured by experiments and estimated theoretically by different authors. The appropriate data are presented in Table 2.6. The table shows that the inflection point of the dependence $\sigma = f(d^{-\frac{1}{2}})$ in a general case forms at $d = 5...50$ nm. The review data in [119] confirm the presence of the inflection point on the dependence $\sigma = f(d^{-\frac{1}{2}})$ (Fig. 2.30] and the range d_{cr} presented in Table 2.6. Evidently, in different materials the inflection point can form at different average grain sizes.

There are also other assumptions regarding the behaviour of the parameter k in the vicinity of the small grain size. According to these assumptions, the transition from the positive and negative value k takes place gradually on reaching several values of d_{cr} (see, for example, Fig. 2.31]. This concept is confirmed by the data for Ni presented in Fig. 2.32. In the review in [122] a wide range of

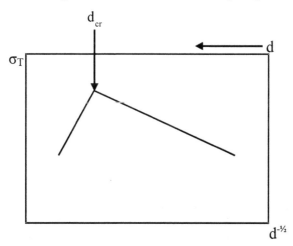

Fig. 2.29. Hypothetical representation of the dependence of the yield point σ_T on the grain size; d_{cr} is the first critical grain size.

Table 2.6. Values of the first critical grain size d_{cr}

d_{cr}, nm	Theory, experiment, review	Authors
5–10	theory	Aust, Erb
10	theory	Carsley et al
10	theory	Aifantis
20–30	theory	Arzt
15	theory	Pande
20	theory	Kim, Estrin, Bush
25	theory	Andrievskii, Glezer
10–50	theory	Conrad
10	experiment, Cu	Sigel
20	experiment, Ni	Yamisava
10–25	review	Ovid'ko, et al.

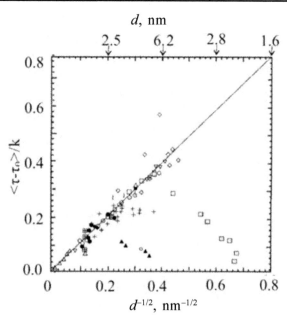

Fig. 2.30. Behaviour of the reduced yield point in dependence of $d^{-1/2}$ for different materials [114].

the grain size is approximated by a single value $k = 0.15$ MPa · m$^{1/2}$ (solid line in Fig. 2.32). However, the figure also shows clearly that the value k decreases with decreasing grain size (see the dotted lines in Fig. 2.32, obtained by the authors). In the grain size range 100...50 nm the value of k decreases to 0.09 MPa · m$^{1/2}$, and then on reaching $d = 25$ nm k becomes equal to 0.06 MPa $\cdot$ m$^{1/2}$. Evidently, the concept in Fig. 2.31 is confirmed by the data presented in Fig. 2.32.

k, MPa$\cdot$m$^{1/2}$

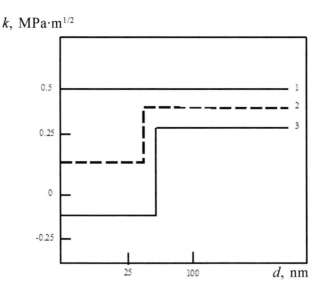

Fig. 2.31. Variance of the behaviour of parameter k in dependence on the grain size d.

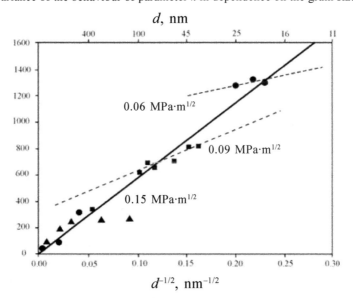

Fig. 2.32. Dependence of the yield point of the grain size $d^{-1/2}$ for nickel according to the data in [1 22]. The dotted lines were obtained by the authors of this book. The values of k for different ranges of the grain size are presented.

The largest number of measurements, carried out on a single metal, was taken for Cu. In these studies, the authors combine the results of a number of investigations. In one case, the existence of the single value $d_{cr} \approx 40$ nm was established (Fig. 2.33]. In another case, where the large number of the data obtained in different measurements were combined (Fig. 2.34) [114], there was a blurred inflection point from the positive value of coefficient $k = 0.10...0.11$ MPa $\cdot$ m$^{1/2}$ to negative values $-0.03...0.07$ MPa $\cdot$ m$^{1/2}$ in the range of the average grain sizes of $d_{cr} \approx 30...12$ nm. If in Fig. 2.34 careful analysis is made of this transition from positive to negative values of k, it may be seen that in the range $d = 70...30$ nm $k = 0.030$ MPa $\cdot$ m$^{1/2}$, and in the range $d = 90...12$ nm $k = 0.015$ MPa $\cdot$ m$^{1/2}$ (see the dotted lines in Fig. 2.34, obtained by the authors of the book). This means that in both Cu and Ni the value of k decreases instead of a sharp transition.

It is evident that for copper there is still no united view regarding the problem of the parameters of the dependence $\sigma = f(d^{-1/2})$. Therefore, in this chapter, to obtain a more detailed review, some of the available data for polycrystalline copper are collected (Fig. 2.35] [9]. Figure 2.35 shows that different authors obtaining different values of k observed always that this value is constant in a wide range of the grain size. However, if the values of k in Fig. 2.35 are averaged out for every narrow grain size range, it may be seen that the value k decreases with a decrease of the grain size. To obtain a more complete pattern of the behaviour of k for Cu, the data presented in Fig. 2.34 and 2.35 were generalised. The final result is presented in Fig. 2.36 [8, 9, 18].

The graph shows that the value of k for Cu decreases, starting at a grain size of 500 μm. The almost constant rate of decrease of k affects four orders of the grain size, from $5 \cdot 10^2$ to $5 \cdot 10^{-2}$ μm. The value of k passes through the well-known average value $k = 0.11$ MPa $\cdot$ m$^{1/2}$ [69] (see the dotted line in Fig. 2.36]. After the value $k = 5 \cdot 10^2$ μm, the value of k rapidly decreases and becomes negative. Figure 2.36 indicates that the first critical grain size d_{cr} is in the range of the values of d slightly lower than 50 nm. Comparison of Fig. 2.36 and Table 2.6 shows that the transition to the negative value of k takes place at a higher value of d_{cr} than the values predicted by the majority of authors systematised in Table 2.6. The experimental data are closest to the prediction by H. Conrad.

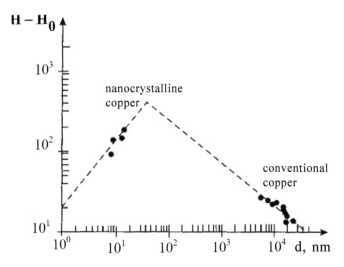

Fig. 2.33. Behaviour of hardness ($H-H_0$) in dependence on the grain size [123].

2.14. Problem of the transition of coefficient k to negative value. The first critical grain size

The displacement of the experimental studies to the region of the microlevel shows directly that the coefficient k depends on the grain size d and, in addition to this, usually decreases with decreasing d. The transition to the negative value of k takes place at the critical grain size.

The most widely encountered variants of the behaviour of parameter k with a decrease of the grain size are: 1) its constancy; 2) jump-like reduction [114, 133] or 3) the change of the sign of k on reaching some grain size [4]. The last case is shown in Fig. 2.29. On reaching the critical grain size d_{cr} an inflection point forms on the dependence $\sigma = f(d^{-1/2})$. The possible dimensions d_{cr} at which this phenomenon occurs are presented in Table 2.6. Initially, this effect was observed in the measuring the microhardness and subsequently at active deformation. This behaviour is explained by the increase of the contribution to the formation of bulk and grain boundary diffusion.

Figure 2.36 shows that there is no sharp inflection point on the $k(d)$ dependence and the variation of k is smooth. The models, describing the variation of coefficient k in the vicinity of the critical grain size are briefly described below. The H–P coefficient increases with increasing average grain size. At large grain sizes the dependence $k = f(d)$ shows a tendency for saturation.

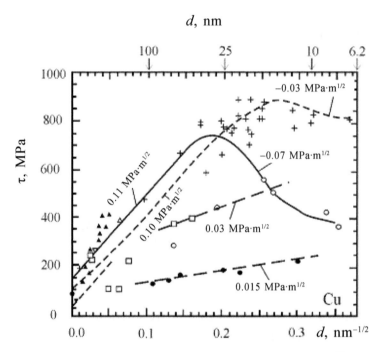

Fig. 2.34. Dependence of the yield point (τ) on the grain size $d^{-1/2}$ for copper [114]. The values of k for different grain size ranges are shown.

Therefore, analysis of Fig. 2.36 and the material presented in this section indicates that the often discussed dependence $k = f(d)$ with a single inflection point does not correspond to the experimental data. In a wide grain size range the value of k decreases with a decrease of the grain size. Of the previously mentioned hypotheses this behaviour is described most efficiently by the scheme proposed in [132, 133], with a large number of inflection points on the dependence of the yield point $\sigma_y = f(d^{-1/2})$.

A large number of researchers have proposed different models for explaining this radical effect. Here it is not possible to examine all these models in detail. Nevertheless, Table 2.7 summarises briefly these concepts. The table gives the authors of the concepts, its physical meaning and the year when it was published.

Table 2.7 shows clearly that initially the authors proposed specific separate ideas for explaining Fig. 2.29. They are very similar. Firstly, it is the model of the low-strength grain and the strong grain boundary region and, vice versa, of the strong grain and the low-strength grain boundary region, or the composite models

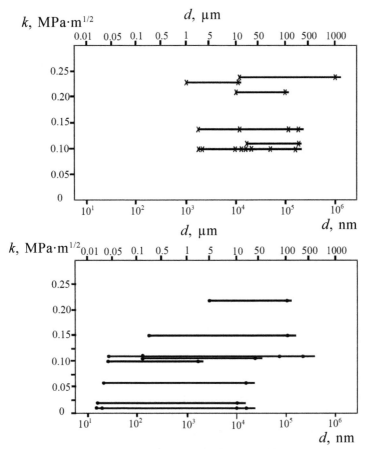

Fig. 2.35. The value of the parameter k for Cu indicating the ranges of the grain size d in which this parameter was measured.

associated with a dislocation structure; secondly, using disclinations for emitting dislocations; thirdly, different structures of the core of the dislocations, lattice and grain boundary and different sliding trajectories; fourthly, different properties of the grain boundaries at the mesolevel and the microlevel and, fifthly, the fact that pile-ups cannot form in the fine grains and, consequently, dislocations cannot be generated from the boundaries of even finer grains. The latter effects are especially important in the integral concept of the behaviour of coefficient k in the region of the nanograins.

After the period 1990–2001, it became clear that the simple models do not include the entire set of the problems associated with the inflection point of the dependence $\sigma = f(d^{-\frac{1}{2}})$ in Fig. 2.29. Therefore, in recent years special attention has been given to the

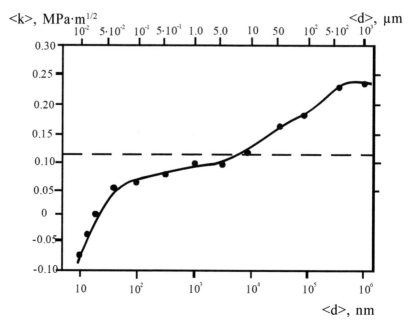

Fig. 2.36. Dependence of the average value of k on the average grain size d (averaging was carried out for the data presented in Figs. 2.27 and 2.28). The dotted line indicates the known average value $k = 0.11$ MPa·m$^{1/2}$.

development of complex models which partially take into account the combined effect of several mechanisms (Table 2.7).

If the data in Table 2/7 are compared with the description of the physical mechanisms, it may be concluded that the problem of the inflection point of the dependence $\sigma = f(d)$ has not been completely solved. Therefore, extensive studies in this area are being continued. The authors believe that attention should be given to the fact that the transition to the negative value of the H–P coefficient differs in most cases from that described in Fig. 2.29. The detailed data, presented in Fig. 2.36, show that the transition of the value of k from positive to negative values is far smoother. Undoubtedly, further studies are required of the relationship of the transition to the negative values of k and physical mechanisms determining this transition.

Some authors, including the authors of the well-known review [95] assume that this behaviour of k is associated with the change of the exponent in equation (2.1). The authors of this book believes that a more accurate interpretation is based on the variation of the H–P coefficient k because there are physical reasons for this. The latest data for the behaviour of parameter k with the variation of the grain size for Fe, Cu, Al, Ni and Ti are presented in Fig. 2.37. They represent the processing and generalisation of several tens of publications which are not listed here. These results have been generalised in the reviews [9, 95, 134]. These data again confirm that the concept of the fast to change of the sign of k from positive to negative at some critical average grain size is not implemented in the experiment, in spite of its attractive nature.

Firstly, Fig. 2.37 shows that the value of k decreases with decreasing grain size. In the case of Al, Ni, Ti and Fe the value

Table 2.7. Concepts for explaining the transition to the negative value of the Hall–Petch coefficient k

No.	Year	Authors of the concept	Physical meaning of the concept
1	1990	Polumbo, Erc, Aust	Dislinations in triple lines
2	1991	Nieh et al	Absence of dislocation pile-ups
3	1992	Valiev	Low-strength grain and strong grain boundary region
4	1992	Scattergood, Koch	Polycrystal as a composition of grains with dislocations inside and without them
5	1993	Bush	Low-strength grain and strong grain boundary region
6	1993	Lu, Sui	Barrier effect from conventional GBs and relay transfer of strain in nanograins
7	1993	Li, Sun, Wang	Soft core in lattice dislocations, Sliding of lattice disloctions in grain body and at GBs
8	1995	Aifantis et al	Low-strength grain and strong grain boundary region
9	1997	Hahn et al.	Soft intergranular interlayers, sliding on them
10	1997	Hahn et al.	Sliding on GBs and migration at GBs

No.	Year	Authors of the concept	Physical meaning of the concept
11	1998	Arzt	Sliding of grain boundary dislocation on GBs
12	1999	Glezer	When the acting stresses become comparable with the sliding stresses on GBs, coefficient k becomes equal to zero
13	2000	Andrievskii,Glezer	
14	2000	Conrad	
15	2001	Gutkin, Ovid'ko, Pande	Complex models taking partially the following mechanisms into account: diffusion on GBs, dislocation and non-dislocation sliding on GBs, emission and absorption of dislocations by GBs, sliding in boundary region, reaction of defects on GBs, accommodation processes
16	2001	Benson, Fu, Meyers	
17	2002	Gutkin, Ovid'ko	
18	2002	Glezer	
19	2003	Nazarov et al.	
20	2003	Sherby et al.	
21	2003	Cheng et al.	
22	2004	Gleiter et al.	
23	2004	Meyers et al.	
24	2004	Mitchell et al.	
25	2004	Li et al	
26	2005	Malygin	

of k approaches zero, and for Cu it becomes even negative. These phenomena occurred in the vicinity of the average grain size of 10 nm which is the first critical grain size. This value is in good correlation with the majority of predictions of the values of d_{cr}, summarised in Table 2.6. It may be concluded that the theoreticians use the situation describing the behaviour of coefficient k. Here, it should be stressed that according to the data obtained in many experiments and studies of the simulation of deformation of the nanocrystals, the transition to the negative value k indicates above all the transition from dislocation mechanisms of deformation to diffusion mechanisms and, in particular, to grain boundary sliding. This process is examined in greater detail in the following sections. Secondly, at a grain size of 50 nm and larger the value of k increases in the following sequence of metals: Al – Cu – Ni – Ti – Fe. This sequence remains unchanged to the value $<d> = 1000$ μm. In other words, the value of k is lower for the FCC metals, higher for HCP metals and the highest for BCC metals. It is necessary to explain whether this sequence is only the property of the crystal lattice or whether as a consequence of

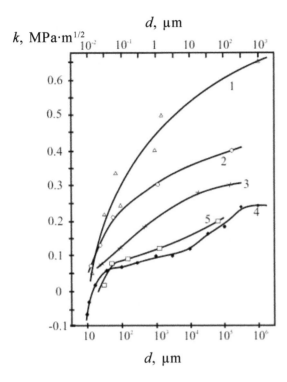

Fig. 2.37. Dependence of the average value of the Hall–Petch parameter <k> on the average grain size d for Fe (1), Ti (2), Ni (3), Cu (4) and Al (5).

different content of the impurities. Thirdly, on approaching the grain size of 10 nm the value of coefficient k rapidly decreases. Its values for different metals converge. This important special feature in the behaviour k, determined to a large extent by grain boundary sliding, requires quantitative explanation. At least, it may be asserted that at a high density of the grain boundaries the properties of different metals be different crystal lattices become very similar. Therefore, it is important to know the list of micromechanisms providing grain boundary sliding.

2.5. Mechanisms of realisation of the Hall–Petch relation at the microlevel

The microlevel for the analysis of the behaviour of the H–P relation deserves special examination. In the grain size 10...100 nm investigations revealed large deviations from the H–P relation and changes of this parameter up to the formation of the negative value of k. Since the decrease of k and, in particular, the transition to the negative value of k prevents the implementation of the main concept – grain refining to increase the yield point – it is natural that a large number of conceptual and theoretical studies have been devoted to get phenomenon. This section will examine gradually the basic problems relating to the behaviour of coefficient k. The main features here are the special features of behaviour during deformation of the UFG and sub-microcrystals with the decrease of the grain size.

The first problem relates to the formation of the dislocation structure. If as a result of grain refining the critical radius R of the Frank–Read source

$$\tau_{\text{F-R}} = \frac{Gb}{R},\tag{2.6}$$

here G is the shear modulus, b is the Burgers vector, becomes comparable with the grain size, the formation of pile-ups by the grain boundaries source is no longer possible and the conditions of formation of the H–P relation at least for the models 1 and 2 (see section 2.12) are violated. The deformation mechanisms and the role of the dislocation components in these mechanisms change. It is clearly evident that the value of the H–P coefficient changes in these conditions.

The second reason for the variation of the behaviour of the H–P relation is a decrease of the linear tension of the dislocations with decreasing grain size. The flow stress decreases correspondingly. Taking into account the linear tension, the H–P coefficient has the following form [135]:

$$k = Gb \ln \frac{d}{r_0},\tag{2.7}$$

where r_0 is the radius of the dislocation core. Equation (2.5) shows clearly that at $d \gg r_0$ the H–P relation is strictly fulfilled. When d approaches r_0 the coefficient k initially decreases and then becomes negative. It is clearly evident that at low values of d the effect of the dislocation mechanisms greatly changes and becomes more complicated.

The third reason, distorting the behaviour of the H–P relation at small grain sizes, is the diffusion mobility through the volume of the grain and at the grain boundaries. In particular, the creep stress as a result of diffusion (Nabarro–Hering creep) has the form [135, 136]:

$$\tau = \frac{\dot{\varepsilon} k T d^2}{C_1 D_v \Omega},$$ (2.8)

where $\dot{\varepsilon}$ is the creep rate, k is the Boltzmann constant, T is the absolute creep temperature, C_1 is a constant, D_v is the coefficient of bulk diffusion, Ω is the atomic volume. It is evident that the stress decreases with decreasing grain size.

If the diffusion mechanisms start to operate, a significant role will be played by the diffusion along the grain boundaries. In particular, the Coble creep stress [136–138]:

$$\tau = \frac{\dot{\varepsilon} k T d^2}{C_2 \delta_b D_{gb} \Omega},$$ (2.9)

where C_2 is a constant, D_{gb} is the coefficient of diffusion along the grain boundaries; δ_b is the thickness of the boundary. Undoubtedly, the Coble creep will reduce the flow stress with a decrease of the grain size in direct proportion with d^3. There are also further three factors, associated with equation (2.7). They will now be discussed. Firstly, the thickness and effectiveness of the grain boundaries increase with the decrease of the grain size. Since δ_b is in the denominator, the stress decreases with increasing δ_b. Secondly, increasing the effectiveness of the GBs creates more suitable conditions for diffusion, because the activation energy decreases (Table 2.8). This results in a corresponding increase of the diffusion coefficient D_{gb} and a decrease of the deformation stress. Thirdly, in addition to grain boundary diffusion, more intensive diffusion takes place along the triple lines. Thus, the operation of the diffusion processes with refining of the grains results in a decrease of the flow stress.

The fourth reason, changing the behaviour of the H–P relation, is grain boundary sliding [94, 139]. Since the grain boundary dislocations are characterised by low values of the Burgers vector b_{gb}, which can be considerably lower than the Burgers vector b of the lattice dislocations ($b_{gb} < b$ or $b_{gb} \ll b$), the stresses of generation of the grain boundary dislocations decrease and dislocation pile-ups appear. In the case of the effect of the grain boundary sliding the flow stresses may also decrease. The high density of the GBs in the UFG and nanomaterials increases the contribution of deformation by grain boundary sliding. Indeed, this is observed in the experiments [8, 94].

Table 2.8. Activation energy of grain boundary processes in UFG copper

Type of process	Activation energy, eV/atom	Reference
Grain boundary diffusion	0.64–0.69	[139]
Coble creep	0.72	[139]
Grain growth	0.70	[94]
Creep of dispersion-hardened material	0.58–0.85	[140]

The fifth reason for complicating the H–P relation is the migration of the grain boundaries (Gleiter) which may start at the very early stages of deformation. Grain by the migration is closely associated not only with the accommodation of deformation processes but also with the processes of emission and absorption of dislocations by the grain boundaries. The set of all the formation of mechanisms, activated by grain refining, naturally changes the behaviour of the H–P relation and reduces the value of k.

A generalisation of the number of deformation mechanisms mentioned here which change or appear at low grain sizes was carried out in [141]. A slightly simplified scheme of the change of the deformation mechanisms in grain refining is shown in Figs. 2.38 [142] and 2.39 [43, 44]. They show that the diffusion processes play a significant role in the strength of the UFG and nanomaterials. In conclusion, it should be stressed that many changes in the behaviour of the H–P relationship in the materials with the ultrafine grain sizes are due to the increasing role of thermally activated processes of plastic deformation.

The sixth reason complicating the behaviour of the H–P coefficient k may operate at any grain size, but is especially important at the microlevel. At the given average grain size the size distribution of the grains has a specific form. Intragranular sliding always starts at the largest grains of the sample, and grain boundary sliding – at the boundaries of the largest grains with most favourable orientation in relation to the external stress [8]. Therefore, the type of grain size distribution also influences the value of the H–P coefficient k. If this factor is taken into account, the values of coefficient k decreases.

2.16. Mechanisms providing contribution to the grain boundary sliding process

This section lists the processes associated with the behaviour of both groups of atoms and the defects that the grain boundaries. Of course, the contribution of each mechanism depends on temperature, the type of stress state, the type of grain boundary and their defective non-equilibrium state. These mechanisms will be listed on the basis of their physical meaning.

I. Diffusion mechanisms: 1) vacancy diffusion; 2) diffusion of atoms and groups of atoms; 3) diffusion displacement of the free or constricted volume; 4) climb of total lattice dislocations, partial lattice dislocations and grain boundary dislocations; 5) migration of the grain boundaries, triple lines and quaternary junctions.

II. Shear mechanisms: 1) sliding of total lattice, partial lattice and grain boundary dislocations; 2) emission of sliding dislocations of different type by the joint disclinations and steps at the grain boundaries; 3) development and collection of stacking faults and anti-phase boundaries; 4) absorption by the grain boundaries of the sliding dislocations of different type.

III. Complex mechanisms: 1) force displacement of the free or constricted volume; 2) splitting and other reactions at triple junctions of the grain; 3) phase transitions in the grain boundary interlayers; 4) phase transitions of the diffusion type in the boundary region; 5) phase transformations in the boundary region of the shear type; 6) formation of waves of directional displacements.

2.17. The number of dislocations in the shear zone and the stress, required for the formation of this zone

At present, it is generally recognised that dislocations in the metallic polycrystals are emitted mostly by the grain boundaries. At least, up to the beginning of the stage III of plastic deformation only the grain boundaries are sources of dislocations. In most cases, the dislocations emitted by the grain boundaries form pile-ups. In the mesopolycrystals, these pile-ups are localised in the vicinity of the game boundaries by which they were emitted. Experiments show that starting with the UFG polycrystals, the pile-ups, emitted by one grain boundary press on the opposite grain boundary. The stress (σ_{gr}) of the formation of a group of n dislocations in a section with length L in the absence of forest dislocations can be expressed by the equation:

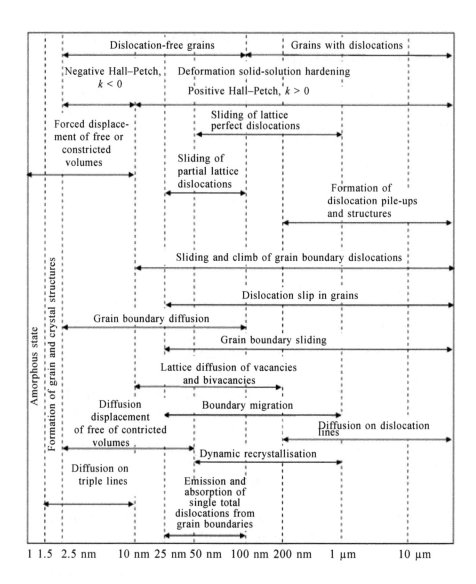

Fig. 2.38. The variation of the dominant deformation mechanisms for copper with a change of the grain size at $T = 300$ K.

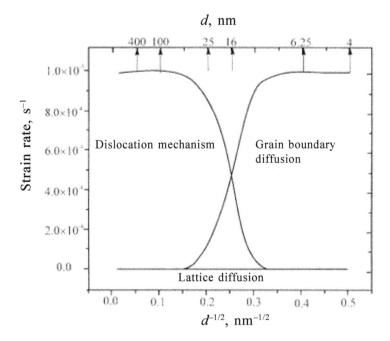

Fig. 2.39. The strain rate determined by the effect of different mechanisms in dependence on the grain size.

$$\sigma_{gr} = \frac{mGbn}{(1-v)\pi L},\qquad(2.10)$$

where m is the orientation factor, G is the shear modulus, b is the Burgers vector, v is the Poisson coefficient.

In emission of the dislocations from the grain boundary the small grain size d restricts the number of emitted dislocations. Until there are several such dislocations, all the models, explaining the H–P effect and based on the formation of a group of dislocations, remain valid. However, when the composition of the groups does not exceed 1–2 dislocations, it is necessary to change the model explaining the mechanism of formation of the yield point. This transition will be examined in greater detail.

The schemes, illustrating in the first case the emission of a group of dislocations by the grain boundary and in the second case the emission of individual dislocations, are shown in Fig. 2.40 [9]. The planar pattern of the process is shown. Figure 2.40a shows the emission of dislocations by a quaternary junction of the grain boundary in one grain and subsequent emission of dislocations, induced by these group of dislocations, in the neighbouring grain. The stress for generation of the dislocations by a source with a radius R is determined by the relationship (2.6). Therefore, two main cases can be observed in dependence on the grain size d: 1) $d \gg R$, pile-ups are generated (Figure 2.40a); 2) $d \approx R$, individual dislocations are generated (Fig. 2.40b). Figure 2.40b shows the case in which there are two such dislocations. The difference between the patterns shown in Figs. 2.40 and 2.40b illustrates the transition from the sliding of groups of dislocations of the sliding of individual dislocations, observed in the fine-grained material. In fact, these schemes describe the transition, in accordance with Table 1.2, from the UFG crystals to nanopolycrystals.

In the works carried out by the authors of the book one of the experimental material was UFG copper, produced by ECAP, and therefore the estimates presented here have been obtained using the data for copper. The dependence $\sigma = f(\varepsilon)$ for UFG copper with the average grain size of $\langle d \rangle = 210$ nm is mostly horizontal (stages VI) at $\sigma = 500$ MPa (Fig. 1.17]. Therefore, estimates are obtained using the relationships (2.6) and (2.10) with the constants $m = 2.2$, G = $5 \cdot 10^4$ MPa, $b = 0.25$ nm and the acting stress of $\sigma = 500$ MPa. In this type of copper dislocation generation starts under the conditions $R = d$ at the grain size $d = 54$ nm. Of course, if there is a stress raiser, this size can become smaller. Formation of groups of two dislocations in accordance with (2.8) takes place at the grain size of $d = 50$ nm. Thus, the scheme in Fig. 2.40b is realised in Cu at $\sigma = 500$ MPa in a grain with the size of $d = 50$ nm. Table 2.9 gives for copper at $\sigma = 500$ MPa the sizes of the grains in which groups of

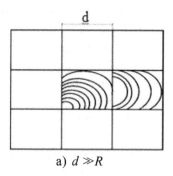

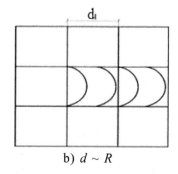

a) $d \gg R$ b) $d \sim R$

Fig. 2.40. The diagrams illustrating the emission of the dislocations by the grain boundaries: a – $d \gg R$, the group of dislocations is emitted by the quaternary junction followed by initiation of the dislocations in the adjacent grain; b – $d \sim R$, the grain boundary emits a small group of dislocations (two groups); d is the grain size, R is the radius of the dislocation source.

dislocations of different power can form. The power of the group is determined by the parameter p – the magnitude of shear

$$p = nd \qquad (2.11)$$

The generation of a group of dislocations from the boundaries of square grains and the formation of a group of the dislocations are shown in Fig. 2.40b. Different dislocation groups are generated ($n \gg 2$) at a large grain size or higher stresses, see Table 2.9.

The pattern of grain boundary sliding, observed in the electron microscope on replicas, is shown in Fig. 2.41. Figure 2.42 shows, for the investigated UFG copper, the dependence of shear on deformation. The magnitude of shear was determined by electron microscopy on replicas separately in the body of the grain (p), sliding of dislocations from boundary to boundary, and separately for sliding along the grain boundaries (P). Figures 2.41 and 2.42 show clearly that the grain boundary sliding is characterised by a high value of shear from the very beginning of deformation. At the same time, the sliding inside the grains in the initial stages of deformation takes place by separate dislocations and small groups which perform shear at $p < 1.5$ nm. The last value is the resolution limit of the replicas. The data in Figs. 2.41 and 2.42 showed that in a grain with the mean size $<d> = 210$ nm of UFG copper sliding initially starts in accordance with Fig. 2.40b – by individual dislocations emitted by one grain boundary and absorbed by the opposite grain boundary (Fig. 2.43). Only at $\varepsilon > 0.3...0.4$ sliding in the medium size grains takes place by the scheme shown in Fig. 2.40b – by groups of dislocations (8 dislocations, Table 2.9). The magnitude of the shear is the resolution limit of the replica, and it is not always possible to observe this type of shear on the surface of the deformed UFG copper.

2.18. Contact stresses. Conventional and accommodation sliding

The deformation of a polycrystalline aggregate consists of the total deformation of the individual grains of the aggregate. Sliding in the individual grains developed mostly in the most heavily loaded sliding systems having the maximum or similar Schmid factor [143, 144]. Usually, there are no more than two such systems in the grain. Since the grains are characterised by different orientation, the anisotropy

Table 2.9. The grain sizes of nanocrystals and UFG copper in which the groups of dislocations of different power form at $\sigma = 500$ MPa, and the magnitude of shear ρ in these grains

d, nm	25	50	100	210	250	500	1000
n	1	2	4	8	10	20	40
p, nm	0.25	0.5	1.0	2.0	2.5	5.0	10

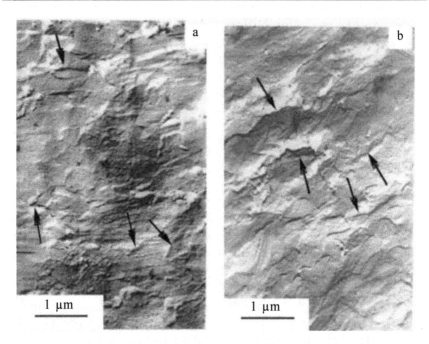

Fig. 2.41. The typical pattern of the surface of the deformed specimen of submicrocrystalline copper observed on the replicas: a – $\varepsilon = 29°$, b – $\varepsilon = 53\%$. The photographs shows coarse traces of sliding at the grain boundaries (indicated by the arrows) and fine traces in the body of the grains.

of elastic models and different dimensions, sources of dislocations of different density act as a boundaries of these grains, and grains at moderate and low temperatures are rigidly bonded together through the boundaries, the deformation of the individual grains becomes incompatible. The latter results in the formation of contact stresses between the grains. This leads to the appearance of accommodation deformation.

If twinning is not possible, two mechanisms of accommodation deformation operate [94]. Firstly, it is intragranular dislocation sliding of the systems with the small or zero Schmid factor (Fig. 2.44). Depending on the depth of penetration of the contact stresses into

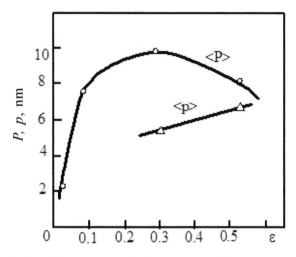

Fig. 2.42. Variation of the magnitude of shear and the distances between the sliding traces with deformation in ultrafine-grained copper. The data are presented for coarse traces at the grain boundaries (P) and the values (p) measured for intragranular sliding.

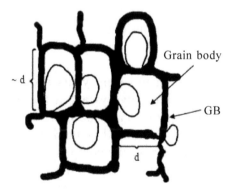

Fig. 2.43. Diagram of sliding of total single dislocations in nanograins: d – the grain size, GB – the grain boundary.

the body of the grain, the accommodation system can be localised either only in the vicinity of the boundary (Fig. 2.44a, the scheme proposed by A. Kochendorfer) [145] or in large areas of the grain (Fig. 2.44b, the scheme proposed by Yu.P. Sharkeev, N.A. Koneva). In the mesopolycrystals, contact stresses usually penetrate into the depth equal to the 1/5...1/10 of the grain size and either the scheme in Fig. 2.44a or 2.44b is realised. Starting with the fine-grained polycrystal, the contact stresses penetrate through the entire grain and the rule of the maximum orientation Schmid factor no longer holds. In this case, sliding systems with any external given Schmid factor can operate. This results in the scheme shown in Fig. 2.44b [146]. Naturally, in the UFG materials and even more so in the sub micro- and nanopolycrystals the contact stresses penetrate through the entire grain. Sliding in each specific grain is determined by the sum of the externally applied and internal contact stresses. Analysis of the sliding pattern in the UFG materials shows [8] that two variants may form. The first variant is based on the action of the main system with the maximum Schmid factor and a slight effect of the secondary systems with the small Schmid factor. The second variant forms when the most developed systems cannot be related only to the systems with the maximum Schmid factor of the external applied stresses. In the latter case, the main systems operate as a result of the presence of contact stresses in the polycrystalline aggregate.

The second mechanism of accommodation sliding is sliding on the grain boundaries. The appropriate illustration of the development of deformation is shown in Fig. 2.45. In the mesopolycrystals, in addition to intragranular sliding with the maximum Schmid factor at moderate temperatures, intergranular sliding develops on the special boundaries similar to twinning boundaries (Fig. 2.45], for example, on the boundaries with the inverse density of the matched sections $\Sigma = 3$.

Experiments carried out in recent years show that the sliding on the grain boundaries in the polycrystals at the mesolevel may take place not only on the special grain boundaries. It appears that the sources of dislocations at the grain boundaries can also operate under one important condition: part of the components of the Burgers vector of the dislocations, emitted from the grain boundary, must be situated in the plane of the boundary [146]. This means that the emission of a group of dislocations from the grain boundary lease automatically to the sliding on these boundaries. This recently discovered effect greatly changes the current views regarding sliding on the high-angle grain boundaries in the mesopolycrystals at moderate temperatures. Thus, the pattern shown in Fig. 2.45 relates not only to the grain boundaries, similar to special boundaries, but also to any other boundary [147].

The grain refinement disrupts the perfection of the grain boundaries. The fraction of the free volume at the grain boundaries increases, the energy of the grain boundaries also increases together with the mobility of the atoms. This is clearly indicated by comparing

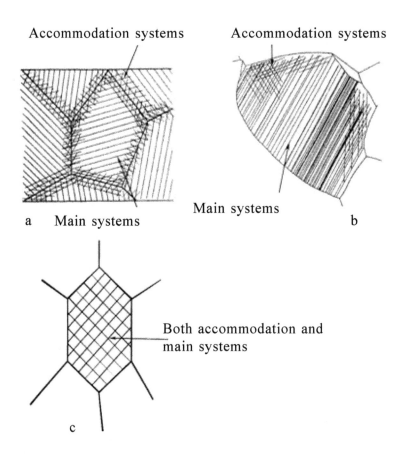

Accommodation systems   Accommodation systems

a   Main systems   Main systems   b

Both accommodation and
main systems

c

Fig. 2.44. Illustration of the main and accommodating sliding: a – accommodation systems local is in the vicinity of the grain boundaries (diagram by A. Kochendorfer); b – accommodation systems occupy a large part of the grain [145]; c – accommodation systems penetrate through the entire grain, being also the main systems [146].

the activation energy of the processes taking place at the grain boundary with different grain sizes [148, 149]. Usually, when the grain size is reduced by an order of magnitude or more the activation energy of the processes taking place at the grain boundaries is greatly reduced. This facilitates the sliding processes at the grain boundaries and increases their contribution to plastic deformation. Experimental studies of the sliding on the grain boundaries in the UFG material (UFG copper) at a low-temperature were described in [8, 94].

With the grain refinement, the contribution to the general deformation of the body of the grains and grain boundaries changes.

Fig. 2.45. Sliding (indicated by the arrows) at the special boundary (the twin boundary) with the inverse density of the coinciding sections $\Sigma = 3$. Cu_3Au alloy, tensile deformation at room temperature.

The contribution of the body of the grains decreases and that of the grain boundaries increases. This becomes especially clear in transition from the UFG polycrystals to nanopolycrystals. At least two sliding mechanisms on the grain boundaries in the UFG polycrystals have been identified. The groups of the grain boundary dislocations slide on the perfect or almost perfect grain boundaries. The imperfect grain boundaries with large distortions of the structure are characterise mostly by the displacement of the free or constricted volume of different type, especially diffusion. The theory of the first type of the processes has not been developed sufficiently, whereas the second theory is being intensively developed.

2.19. Conclusion

The investigation of the problems associated with the hardening mechanism of the UFG materials is being completed. It is important to mention several results. Firstly, it should be noted that the UFG materials are characterised by a high yield point. All the hardening mechanisms, which operate in the conventional polycrystals at the mesolevel, also operate in the polycrystals at the microlevel. The high density of the grain boundaries creates additional resistance to the movement of dislocations and increases the yield point of the UFG materials. To this it is necessary to add the high initial density of dislocations distributed at the grain boundaries and at their junctions. The high level of the internal stress fields in the UFG materials also increases the yield point. After reaching the critical grain size $d = 100$ nm the dislocation density in the grains decreases butter the density of partial and contact disclinations increases and the high level of the internal stresses is maintained. The type of mechanism changes with further grain refinement – the formation of pile-ups becomes more complicated, the intensity of sliding along the grain boundaries increases, and the diffusion processes and migration of the grain boundaries are activated. The Hall–Petch relation is fulfilled prior to reaching the critical grain size $d \approx 20...$ 50 nm. When the average grain size becomes equal to this critical size, the Hall–Petch coefficient becomes negative and grain boundary hardening is replaced by grain boundary softening. At d = 50 nm and lower the strength (deal strength) will not increase further for these reasons with grain refinement and can in fact decrease.

Actually, the first ($d = 20...50$ nm) and second ($d \approx 100$ nm) critical grain sizes reduce the role of the deformation mechanism by lattice dislocations and result in the gradual replacement of this mechanism by the sliding of grain boundary dislocations, diffusion processes and the forced displacement of the free and constricted volume. The relationships governing the deformation of the crystalline material become similar to the relationships of the formation of the amorphous material.

References

1.   E.O. Hall, *Proc. Phys. Society*, 1951, V. 64B, 747–753.
2.   N.J. Petch, *J. Iron Steel Inst.*, 1953, V. 174, 25–28.
3.   J.P. Hirth, *Met.Trans.*, 1972, V. 3, 3047–3067.
4   N. Hansen, *Met.Trans.*, 1985, V. 16A, 2167–2190.
5.   V.S. Ivanova, et al., The role of dislocations in the hardening and fracture of metals. Moscow, Nauka, 1965.
6.   V.I. Trefilov, et al., Deformation hardening and fracture of polycrystalline metals (ed. V.I. Trefilov), Kiev, Naukova Dumka, 1989.
7.   L.I. Tushinsky, Structural theory of structural strength of materials, Novosibirsk, Publishing House of the National Technical University, 2004.
8.   E.V. Kozlov, et al., *Fiz. mezomekhanika*, 2004, V. 7. No. 4, 93–113.
9.   E.V. Kozlov, ET AL., *Fiz. mezomekhanika*, 2006, V. 9, No. 3, P. 81–92.
10. E.V. Kozlov, et al., *Fiz. mezomekhanika*, 2007, V. 10, No. 3, 81–92.
11. E.V. Kozlov, et al., Zf. Funktional. Mater., 2007, V. 1, No. 1, 21–24.
12. D. Wolf, K.L. Merkle, Correlation between the structure and energy of grain boundaries in metals. Atomic-level structure and properties (eds. D. Wolf and Y. Sidney).

London, Chapman and Hall, 1992, 87–150.
13. M.J. Mills, *Mater. Sci. Eng.*, 1993, V. 166, 35–50.
14. Yield, flow and fracture of polycrystals (ed. T. N. Baker), London and New York, Appl. Sci. Publishers, 1983.
15. Grain size and mechanical properties – fundamentals and applications, V. 362. (eds M.A. Otooni, R.W. Armstrong, N.J. Grant and K. Ishizaki), Pittsburgh, Mat. Res. Society, 1995 270p.
16. V.M. Segal, et al., Processes of plastic structure formation in metals, Minsk, Nauka i tekhnika, 1994.
17. R.Z. Valiev, I.V. Alexandrov. Nanostructured materials obtained by severe plastic deformation, Moscow, Logos, 2000..
18. N.A. Koneva, et al., *Izv. RAN, Ser. fizicheskaya*, 2006, V. 70, No. 4, 582–585.
19. E.V. Kozlov, et al., *Vestnik Tamboskogo Univ.*, 2003, V. 8, No. 4, 509–513.
20. E.V. Kozlov, et al., *Izv. RAN, Ser. fizicheskaya*, 2009, V. 73, No. 9, 1295–1301.
21. R.Z. Valiev, I.V. Alexandrov, Volumetric nanostructured metallic materials, Moscow, Akademkniga, 2007.
22. T.G. Langdon, et al., *IOM*, 2000, V. 52, No. 4, 30–33.
23. R.Z. Valiev, T.G. Langdon, *Progr. Mat. Sci.*, 2006, V. 51, 881–981.
24. P.L. Sun, et al., *Mat. Sci. Eng.*, A, 2000, V. 283, 82–85.
25. G.I. Raab, et al., *Mat. Sci. Eng.*, A, 2004, V. 382, 30–34.
26. K. Matsubara, et al., *Acta Mat.*, 2003, V. 51, 3073–3084.
27. I.-Y. Suh, et al., *Scr. Mat.*, 2003, V. 49, 185–190.
28. I. Alkorta, et al., *Scr. Mat.*, 2002, V. 47, 13–18.
29. A.P. Zhilyaev, et al., *Acta Mat.*, 2003, V. 51, 753–765.
30. R.Z. Valiev, et al., *Progress in Materials Science*, 2000, V. 45, 103–189.
31. V.V. Gubernatorov, et al., *Fiz. Met. Metalloved.*, 2004, V. 98, No. 4, 83–87.
32. D.A. Rigney, et al., *Scr. Mat.*, 1992. V. 27. 975-980.
33. Y. Sato, et al., *Scr. Mat.*, 2001, V. 45, 109–114.
34. Ultrafine grains in metals (ed. by L.K. Gordienko). Moscow, Metallurgiya, 1973.
35. E.V. Kozlov, *Voprosy materialovedeniya*, 2002, No. 29 (1), 50–69.
36. M.A. Meyers, *Progress in Materials Science*, 2006, V. 51, 427–556.
37. R.Z. Valiev, et al., *Progress in Materials Science*, 2000, V. 45, 103–187.
38. D.B. Witkin, E.J. Lavernia, *Progress in Materials Science*, 2006. V. 51. 1-60.
39. B.Q. Han, et al., *Rev. Adv. Mater. Sci.*, 2005, No. 9, 1–16.
40. R.A. Andrievsky, A.M. Glezer, *Usp. Fiz. Nauk*, 2009, V. 179, No. 4, 337–358.
41. Ch.V. Kopetsky, et al., The grain boundaries in pure materials. Moscow, Nauka, 1987.
42. V. Randle, *Acta Met. Mater.*, 1994, V. 42, No. 6, 1769–1784.
43. Kim H.S., et al., *Acta Mater.*, 2000, V. 48, 493–504.
44. Kim H.S., et al., *Mater. Sci. Eng.*, 2001, V. A316, 195–199.
45. O.B. Perevalova, *Fiz. Met. Metalloved.*, 1999, V. 88, No. 6, 68–76.
46. E.V. Konovalova, et al., *Metallofizika i noveishie tekhnologii*, 2001, V. 23, No. 5. 655–670.
47. Koneva N.A., et al., *Mat. Sci. Forum*, 2008, V. 584–586, 269–274.
48. Romanov A.E., Vladimirov V.I. Disclinations in crystalline solids. Dislocations and disclinations (ed. FN.R. Nabarro), V. 9, Amsterdam–Tokyo, Elsevier, 1992, 191–250.
49. Palumbo G., et al., *Scr. Met. Mater.*, 1990, V. 24, 2347–2350.
50. Valiev R.Z., Aleksandrov I.V., Nanostructured materials obtained by severe plastic deformation, Mosco, Logos, 2000.

51. Gutkin M.Yu., Ovid'ko I.A. Physical mechanics of deformable nanostructures, V. I. Nanocrystalline materials, St. Petersburg, Publishing house of the IPM RAS, 2003.
52. Gutkin M.Yu., Ovid'ko I.A., *Rev. Adv. Mater. Sci.*, 2003, V. 4, 79–113.
53. Kozlov E.V., et al., *Ann. Chim. Fr.,* 1996, V. 21, 427–442.
54. Koneva N.A., et al. Types of grains and boundaries, joint disclinations and dislocation structures of SPD-produced UFG materials, in: Nanomaterials by severe plastic deformation. (Ed. by M.J. Zehetbauer, R.Z. Valiev), Weinheim, Wiley VCH, 2004. 357–362.
55. Fedorov A.A., et al., *Scr. Mat.*, 2002, V. 47, 51–55.
56. Gutkin M.Yu., Ovid'ko I.A., *Phil. Mag.*, 2004, V. 84, No. 9, 847–863.
57. Ovid'ko I.A., Sheinerman A.G., *Acta Mat.*, 2004, V. 52, 1201–1209.
58. Valiev R.Z., et al., *Metallofizika*, 1992, V. 14, No. 2, 58–62.
59. Lazarenko A.S., et al., *Metallofizika,* 1991, V. 13, No. 3, 26–33.
60. Koneva N.A., et al., in: Structure, phase transformations and properties of nanocrystalline alloys, (ed by N.I. Noskova), Ekaterinburg, IFM UrB RAS, 1997, 125–140.
61. Koneva N.A., in: Severe plastic deformation. Toward bulk production of nanostructured materials, (ed. B.S. Altan), New York, Nova Science Publishers Inc., 2005, 249–274.
62. Zhilyaev A.P., et al., *Acta Mat.*, 2003, V. 51, 753–765.
63. Noskova N.I., Mulyukov R.R., Submicrocrystalline and nanocrystalline metals and alloys, Ekaterinburg, UrO RAN, 2003.
64. Koneva N.A., et al., in: New methods in physics and mechanics of deformable solids. Part I, (ed. by VE Panin), Tomsk, TSU, 1990, 83–93.
65. Kozlov E.V., et al., in: Ultrafine grained materials, II, USA, TMS, 2002, 419–428.
66. Kozlov E.V., et al., *Deformatsiya i razrushenie materialov*, 2009, No. 6, 22–27.
67. Kozlov E.V., et al., *Mat. Sci. Eng.*, 2004, V. A387–389, 789–794.
68. Perevalova O.B., et al., *Fiz. Met. Metalloved.,* 2004, V. 98, No. 5, 78–84.
69. Meyers M.A., Chawla K.K., Mechanical behavior of materials, New York, Prentice Hall, 1999.
70. Trefilov V.I., et al., Deformation hardening and fracture of polycrystalline metals, Kiev, Naukova Dumka, 1987.
71. Kocks U.F., *Met. Trans.*, 1970, V. 1, No. 5, 1121–1143.
72. Hirth J.P., *Met. Trans.,* 1972, Vol. 3, 3047–3067.
73. Gleiter G., Chalmers B., Large-angle grain boundaries, translation from English, Moscow, Mir, 1975.
74. Kaibyshev O.A., Valiev R.Z., Grain boundaries and the properties of metals, Moscow, Metallurgiya, 1987.
75. Starostenkov M.D., et al., *Acta Metallurgica. Sinica* (English Letters), 2000, V. 13, No. 2, 540–545.
76. Takasugi T., Structure of grain boundaries // Intermetallic compounds. V.I. Principles. (Eds, J. H. Westbrook and R. L. Fleischer). London, John Wiley and Sons, 1994. 585-607.
77. Mughrabi H., et al., *Phil. Mag.* A, 1986, V. 53, No. 6. 793–813.
78. Mughrabi H., *Mater. Sci. and Eng.,* 1987, V. 85, 15–31.
79. Kozlov E.V., Koneva N.A., *Izv. VUZ Fizika*, Supplement, 2002, V. 45, No. 3, 52–71.
80. Ashby M.F., *Phil. Mag.*, 1970, V. 21, No. 170, 399–424.
81. Ashby M.F., The deformation of plastically non-homogeneous alloys, in: Strengthening methods in crystals, London, Science Publishers, 1971, 137–190.
82. Sharkeev Yu.P., et al., Scheme of development of sliding in grains of polycrystals

with a fcc lattice, *Fiz. Met. Metalloved.*, 1985, Vol. 60, No. 4, 815–821.

83.   Thompson A.W., et al., *Acta Met.,* 1973, V. 21, 1017–1028.

84.   Murr L.E., *Met. Trans.* A, 1975, Vol. 6A, 505–513.

85.   Meyers M.A., et al., *Mat. Sci. Eng.,* 2002, Vol. A322, 194–216.

86.   Koneva N.A., et al., in: Physics of defects in surface layers of materials, ed. A.E. Romanov, Leningrad, AF Ioffe Institute, 1989, 113–131.

87.   Kozlov E.V., et al., in: Disclination and rotational deformation of solids (ed. A.E. Romanov) Leningrad, AF Ioffe Institute, 1990, 89–125.

88.   Koneva N.A., Kozlov E.V., in: Structural levels of plastic deformation and fracture (ed. by V.E. Panin), Novosibirsk, Nauka, Siberian Branch, 1990, 123Leningrad, AF Ioffe Institute186.

89.   Koneva N.A., et al., in: Investigations and applications of severe plastic deformation (eds T. C. Lowe and R.Z. Valiev), NATO Science Series 3, High Technology, 2000. V. 80, 121–126.

90.   Koneva N.A., et al., *Izv. AN, Ser. Fizicheskaya*, 1998, V. 62, No. 7, 1350–1356.

91.   Kozlov E.V., et al., in: Ultrafine grained materials II, (eds. Y.T. Zhu, TG Langdon, R.Z. Mishra, et al), USA, TMS Publication, 2002, 419–428.

92.   Nasarov A.A., et al., *Nanostructured Materials*, 1994, V. 4, No. 1, 93–101.

93.   Kozlov E.V., et al., *Mat. Sci. Eng.*, 2004, Vol. A387–389, 789–794.

94.   Valiev R.Z., et al., *Acta Met. Mater.*, 1994, V. 42, No. 7, 2467–2475.

95.   Meyers M.A., et al., *Progr. Mat. Sci.*, 2006, V. 51, 427–556.

96.   Alexandrov I.V., Enikeev N.A., *Mater. Sci. Eng.*, 2000, V. A286, 110–114.

97.   Valiev R.Z., Musalimov R.Sh., *Fiz. Met. Metalloved.*, 1994, Vol. 78, No. 6, 114–121.

98.   Palumbo G., et al., *Scr. Met.*, 1990, V. 24., 2347–2350.

99.   Suryanarayana C., et al., *J. Mater. Res.*, 1992, V. 7, No. 8, 2114–2128.

100. Wang N., et al., *Acta Met. Mater.*, 1995, V. 43, No. 2, 519–528.

101. Kozlov E.V., et al., *Izv. RAN, Ser. Fiz.*, 2009, 73, No. 9, 1295–1301.

102. Kozlov E.V., et al., *Mat. Sci. Forum*, 2008, V. 584–586, 33–40.

103. Dick von E., *Z. Metallkde*, 1970, V. 61, 451–454.

104. Ono N., Karashima S., *Scr. Met.*, 1982, V. 16, 381–384.

105. Gertsman V.Y., et al., *Acta Met. Mater.*, 1994, V. 42, No. 10, 3539–3544.

106. Courtney T.H., Mechanical behavior of materials, Singapore, McGraw Hill International Editions, 2000..

107. Armstrong R.W., Douthwaite R.M., *Mater. Res. Soc. Symp. Proceed.*, 1995, V. 362, 41–47.

108. Hansen N., Ralph B., *Acta Met.,* 1982, V. 30, 411–417.

109. Thompson A.W., Backofen W.A., et al., *Met. Trans.*, 1971, V. 2, 2004–2051.

110. Gray G.T. III, et al., *Nanostructured Materials*, 1997, V. 9, 477–480.

111. Johnston T.N., Feltner C.E., *Met. Trans.*, 1970, V. 1, 1161–1167.

112. Sinclair C.W., Poole W.J., in: Ultrafine Grained Materials, III, Warrendale, TMS, 2004, 59–64.

113. Tabachnikova E.D., et al., in: Structure and properties of nanocrystalline materials (ed. by G.G. Taluts and N.I. Noskova), Ekaterinburg, UrB RAS, 1999, 103–107.

113. Tabachnikova E.D., et al., in: Structure and properties of nanocrystalline materials. (ed. by G.G. Taluts and N.I. Noskova), Ekaterinburg, UrB RAS, 1999, 103–107.

114. Masumura R.A., et al., *Acta Mater.*, 1998, V. 46, No. 13, 4527–4534.

115. Iger R., et al., *Mater. Sci. Eng.*, 1999, V. A264, 210–214.

116. Sanders P.G., et al., *Acta mater.*, 1997, V. 45, No. 10, 4019–4025.

117. Neiman G.W., et al., *Nanostructured materials*, 1992, V. 1, 185–190.

118. Suryanarayanan R., et al., *J. Mater. Res.*, 1996, V. 11, No. 2, 439–442.

119. Weertman J.R., et al., *MRS Bulletin*, 1999, No. 24, 44–50.
120. Cottrell A.H., Dislocations and plastic flow in crystals, Moscow, Metallurgizdat, 1958.
121. Armstrong R., et al., *Phil. Mag.*, 1962, V. 7, No. 73, 45–58.
122. Li J.C.M., *Trans. AIME*, 1963, V. 227, 239–247.
123. Conrad H., in: Ultrafine grain in metals, Moscow, Metallurgiya, 1973, 206–219.
124. Meaking J.D., Petch N.J., *Phil. Mag.*, 1974, V. 30, 1149–1158.
125. Orlov A.N., *Fiz. Met. Metalloved.*, 1977, 44, No. 5, 966–970.
126. Li J.C.M., *J. Austral Inst. Metals*, 1963, V. 8, 206–212.
127. Meyers M.A., *Phil. Mag. A*, 1982, V. 46, No. 5, 737–759.
128. Trefilov V.I., *Dokl. Akad. Nauk SSSR*, 1985, 285, No. 2, 109–112.
129. Panin V.E., Yelsukova T.F., in: Structural levels of plastic deformation and fracture. (ed. by VE Panin), Novosibirsk, Nauka, Siberian Branch, 1990, 77–123.
130. Xiao X., et al., *Mat. Sci. Eng.*, 2001, V. A301, 35–43.
131. Artz E., *Acta Mater.*, 1998, V. 46, No. 16, 5611–5626.
132. Liu X.D., et al., *Mater. Trans. JIM*, 1997, V. 38, No. 12, 1033–1039.
133. Sanders P.O., et al., *Mater Sci. Eng.*, 1997, V. A234–236, 77–82.
134. Within D.B., Lavernia E.J., *Progr. Mat. Sci.*, 2006, V. 51, 1–60.
135. Arzt E., *Acta Mater.*, 1998, V. 46, No. 16, 5611–5626.
136. Mohamed F.A., Li Y., *Mat. Sci. Eng.*, 2001, V. A298, 1–15.
137. Arzt E., et al., *Acta Met.*, 1983, V. 31, No. 12, 1977–1989.
138. Ogino Y., *Scripta Mater.*, 2000, V. 42, 111–115.
139. Cai V., et al., *Mater. Sci. Eng.*, 2000, V. A286, 188–192.
140. Shen B.L., et al., *Scr. Mater.*, 2000, V. 41, 893–898.
141. Cheng S.,et al., *Acta Mater.*, 2003, V. 51, 4505–4518.
142. Kozlov E.V., et al., Fiz. mesomekhanika, 2011, V.14, No. 3, 95–110.
143. Honeycombe R.W., Plastic deformation of metals [Russian translation], Moscow Mir, 1972.
144. Backofen W.A., Deformation processes, Moscow, Metallurgiya, 1977.
145. Sharkeev Yu.P., et al, *Fiz. Met. Metalloved.*, 1985, V. 60, No. 4, 815–821.
146. Perevalova O.B., Koneva N.A., *ibid*, 2003, V. 98, No. 4, 106–112.
147. Popova N.A., Lapsker I.A., in: Plastic deformation of alloys (ed. L.E. Popov and N. A. Koneva), Tomsk, TSU, 1986, 241–248.
148. Monzen R., Sumi Y., *Phil. Mag. A*, 1994, V. 70, No. 5, 805–817.
149. Li Y.J., et al., *Acta Mater.*, 2004, V. 52, 5009–5018.

Main components of the dislocation structure and the role of the dimensional factor

3.1. Problem of classification of dislocation structure components

3.1.1. Components of the dislocation structure

Over many years, the dislocation structure has been characterised by the dislocation density ρ. The development of dislocation science has resulted in the sub-division of the value of ρ to components with different physical meaning. The scalar density of dislocations can be divided to the density of the moving dislocations ρ_m and the density of stopping dislocations ρ_{st} [1]. Naturally

$$\rho = \rho_m + \rho_{st}. \tag{3.1}$$

On the other hand, the scalar density of dislocations consists of the dislocations moving in the active slip plane ρ_m and the forest dislocations ρ_f which do not belong to any active slip plane [2]. Now:

$$\rho = \rho_m + \rho_f \tag{3.2}$$

The following division of the scalar dislocation density ρ into two components was more precise. Firstly, the dislocations, accumulated in the volume of the material, are initially emitted by their sources and then are inhibited as a result of reactions with other dislocations. Both the multiplication of the dislocations and the reactions are random processes. Therefore, this group of dislocations is regarded as statistically stored ρ_s [3]. The statistically stored dislocations are inhibited by relatively weak barriers – other dislocations. If the material contains stronger barriers – second phase particles and the grain boundaries – plastic deformation gradients form. If such gradients are present, then in addition to the dislocation density ρ_s there is also the storage of the geometrically necessary dislocations (GND) with the density ρ_G [3]. In this case

$$\rho = \rho_s + \rho_G. \tag{3.3}$$

The presence of geometrically necessary dislocations is often associated with the distortion of the crystal lattice [4].

There is also the division of the dislocations natural for the dislocation theory into positively charged (ρ_+) and negatively charged (ρ_-) [5]. The sum of these dislocations gives the total scalar dislocation density

$$\rho = \rho_+ + \rho_-. \tag{3.4}$$

and their difference gives the excess dislocation density $\rho_\pm$ [6, 7]:

$$\rho_\pm = \rho_+ + \rho_-. \tag{3.5}$$

The excess dislocation density is directly linked with the curvature –torsion of the crystal lattice:

$$\rho_\pm = \frac{1}{b}\frac{\partial\varphi}{\partial\ell} = \frac{\chi}{b} = (Rb)^{-1}, \tag{3.6}$$

where b is the Burgers vector, φ is the angle of inclination of the crystallographic plane, ℓ is the distance on the plane, $\dfrac{\partial\varphi}{\partial\ell}$ is the gradient of the curvature–torsion of the crystal lattice, χ is the curvature – torsion of the crystal lattice, R is the curvature radius of the crystal.

3.1.2. Strain gradient, the density of geometrically necessary and excess dislocations

The density of the geometrically necessary dislocations can be expressed by the strain gradient [4, 8]:

$$\rho_G = (Rb)^{-1}. \tag{3.7}$$

Comparison of (3.6) and (3.7) shows that

$$\rho_G = \rho_\pm. \tag{3.8}$$

The excess dislocation density is equal to the density of the geometrically necessary dislocations. The latter are the stored dislocations which are required for the accommodation of the curvature of the crystal lattice formed as a result of the heterogeneity of plastic deformation, i.e., as a result of the presence of the strain gradient [9]. Distortion of the crystal lattice in the vicinity of the grain boundaries can represented both in units of $\rho_\pm$, and in units of χ [10–12].

3.1.3. Grain size and the density of geometrically necessary dislocations

To describe the work hardening of the polycrystalline aggregate it has been attempted to link the density of geometrically necessary dislocations ρ_G with the average grain size (d).

In [3, 4] it was proposed that:

$$\rho_G = \frac{\varepsilon}{4bd}, \tag{3.9}$$

where ε is the strain. This scheme is satisfied by the Conrad model [13] in which the total dislocation density ρ is inversely proportional to the mean grain size d:

$$\rho \approx \frac{\varepsilon}{0.4bd}. \tag{3.10}$$

The equations (3.9) and (3.10) coincides with the accuracy to the coefficient and, therefore, $\rho_s > \rho_G$ (approximately by an order of magnitude). This means that at the conventional grain dimensions at the mesolevel $\rho > \rho_G$. For the nanograins the relationship can be inverse. The theoretical estimate of the coefficients in (3.9) and (3.10) is not strict and, therefore, this problem must be investigated further.

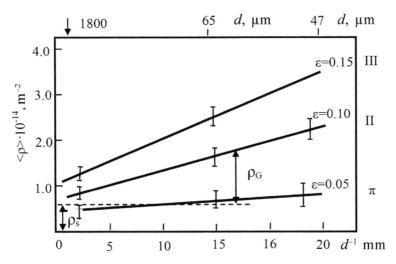

Fig. 3.1. The dependence of dislocation density on the inverse average grain size. Ni$_3$Fe alloy (misordered state). The deformation stages are given on the right. Deformation – tensile loading at T_{room}.

3.1.4. Methods of measuring the density of geometrically necessary dislocations

The equations (3.9) and (3.10) indicate that both ρ and ρ_G and $\rho_\pm$ are inversely proportional to the grain size and depend on the strain. This was confirmed by experiments in [10, 14]. Thus, the first method of determination of ρ_G is based on using the dependence $\rho = f(d^{-1})$ [10] (Fig. 3.1). For the conventional sizes of the mesograins [15], 40... 450 μm, since in most cases $\rho_s > \rho_G$ (Fig. 3.2).

The second method is based on equation (3.8), i.e., on the equality of ρ_G and $\rho_\pm$. The method of measuring $\rho_\pm$ was described in detail in the studies by the authors (see, for example [6, 7, 11]]. To determine $\rho_\pm$ it is necessary to measure the parameters of extinction contours [6, 7] (Fig. 3.3).

The third method is based on measuring the dislocation density ρ as a function of the distance from the grain boundary (Fig. 3.4). At the grain boundaries $\rho = \rho_G + \rho_s$, in the body of the grain $\rho \approx \rho_s$. Measuring the difference of the dislocation densities at the grain boundary and in the centre of the grain it is possible to determine the values of ρ_G. Since the dislocation densities at the grain boundary and in the body of the grain always differ, measuring ρ_s and ρ_G for different strains one can obtain the dependence on

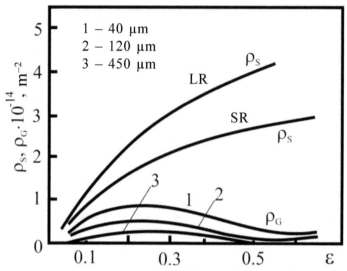

Fig. 3.2. Dependence of ρ_G and ρ_s on the strain for the Ni_3Fe alloy in the conditions with the long-range (LR) and short-range (SR) atomic order. The graph shows the grain sizes for which the values of ρ_G and ρ_s were determined (ρ_G for the alloy in the conditions with the LR).

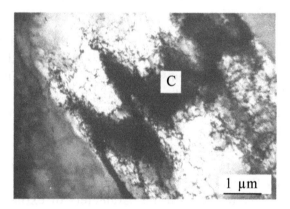

Fig. 3.3. Deformation extinction contour (C) in the Cu +6 at.% Mn alloy.

the strain of both ρ_s and ρ_G. The results of these measurements are presented in Fig. 3.5.

The fourth method is based on the measurement of the parameters of the shear zones – main, secondary and accommodation sliding systems [16]. Figure 3.6 shows the diagram of the sliding system in the grain of a polycrystal. The graph shows the primary system, the secondary system and the accommodation systems situated at the grain boundaries. The primary and secondary systems provide a

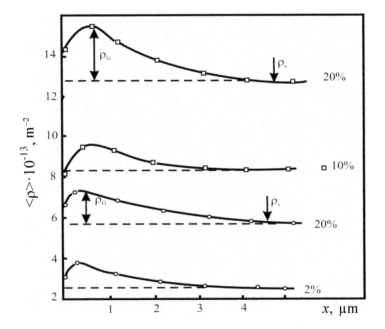

Fig. 3.4. Dependence of dislocation density on the distance from the grain boundary at different strains: o – Cu +0.5 at.% Al, □– Cu +10 at.% Al.

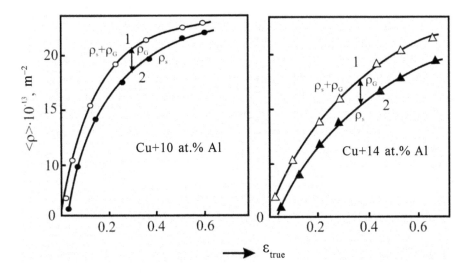

Fig. 3.5. Scalar dislocation density in the body of the grain ρ_s and in the vicinity of the grain boundary $\rho_s + \rho_G$.

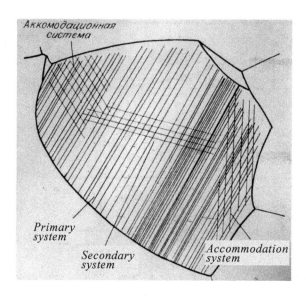

Fig. 3.6. Diagram of the grain of the polycrystal [16].

contribution to ρ_s and the accommodation systems to ρ_G. Using the data on the magnitude of shear and the density of the slip traces, it is possible to separate the contributions of ρ_s and ρ_G to the dislocation density ρ. In transition to the ultrafine-grained and nanocrystals it is important to separate intragranular and grain boundary sliding because the latter is usually not of the dislocation type [15, 17, 18].

Thus, this procedure can be used to analyse the contributions to the scalar density. The density of the geometrically necessary dislocations and the density of excess dislocations are linked by the relationship $\rho_G = \rho_\pm$. The geometrically necessary dislocations form in the polycrystalline material. The density of these dislocations depends on the structure of the grains of the polycrystal [19].

3.2. The scalar density of dislocations in dislocation fragments with different types of substructure

The problem of storage of dislocations in the deformed materials is still complicated and has not been solved. This claim applies to both the investigations of pure metals and solid solutions with a relatively simple dislocation substructure and the materials with a complicated substructure. In these materials phase transformations take place prior to or during plastic deformation [20, 21]. A typical example are the substructures formed in martensitic steels in which in addition to the grains there are also dislocation cells, fragments, packets and laths. Similar structures in which some boundaries are situated inside the other substructural formations can form during plastic deformation in the ultrafine-grained polycrystals, especially in the conditions of dynamic recrystallisation and in other processes [22, 23].

In the microrange of the dimensions of the grains (d) and fragments (d_{fr}) there are the relationships in the storage of dislocations in comparison with the relationships in the mesorange. This phenomenon is characteristic for both ultrafine-grained polycrystals and for small fragments, found in the deformed martensitic steels [24–28]. Quantitative studies by transmission electron microscopy showed some analytical dependences of the scalar dislocation density (ρ) on the size of the grains and the fragments. It is well known that the relationship between the sizes of the grains, dislocation cells and fragments and the dislocation density plays an important role in the theory of dislocation substructures and the dislocation hardening concepts [29,30].

This section describes the results of investigation by transmission electron microscopy of the evolution of the dislocation substructure with the measurement of the scalar density of dislocations in a martensitic steel. It also determines the relationship of the scalar density of dislocations with the grain sizes, the size of the dislocation fragments and the cells. The investigations were carried out on a martensitic steel containing 0.34% C+0.40% Ni+0.30% Cr+0.60% V+0.6% Mo, the balance – Fe. The specimens were quenched from 1000°C, followed by tempering at $T = 600°C$, $t = 6.5$ hours. The specimens produced from the tempered steel were deformed by tensile loading to different strains in the range 5...90%. The deformed specimens were machined in an electrospark machine to produce sheets with a thickness of 0.1 mm which were subsequently thinned by electropolishing to the required thickness for examination in the electron microscope. The foils, prepared by this procedure, were examined in an EM-125K electron microscope, fitted with a goniometer, at an accelerating voltage of 125 kV. The resultant electron microscopic images were used for the identification of the type of substructures produced in the investigated steel, and the secant method was used to measure the scalar density of dislocations, both the mean value in the volume of the material and in different components of the substructure. Special attention was given to the dislocation fragments with a different type of substructure in them.

The statistical processing of the results was carried out in continuous sections of the specimens with the area of ~80 μm², containing 500–1000 dislocation fragments, observed in the steel.

3.2.1. Dependence of the dislocation density on the grain size in ultrafine-grained polycrystals

The dependences $\rho = f(d)$ are determined by the type of dislocation structure formed in the ultrafine grains of the metals. The studies [24–28] investigated the dependence of the dislocation density on the grain size of the ultrafine grained polycrystals of copper and nickel in the conditions of formation of the network or cellular (or fragmented) substructure in them. The results, published in [24–28], indicate that the main relationships, linking the dislocation density ρ and the grain size d at the microlevel for pure metals, has the form

$$\rho = Cd, \qquad (3.11)$$

where C is a constant. The relationship (3.11) is fulfilled in the microregion of the grain size of the metals. The method of production of the ultrafine polycrystals, ECAP or THP (ECAP – equal channel angular pressing, THP – torsion under hydrostatic pressure) does not change these relationships. The formation inside the grains of the cellular, fragmented or network substructure in the pure ultrafine metals has no effect on the relationship (3.11) and determines only the value of the constant C. Fulfilment of the relationship (3.11) indicates that in the microrange of the grain sizes a decrease of d decreases the scalar density of dislocations. When the size $d \approx 100$ nm is reached, the grains become dislocation-free. This determines the value of the second critical grain size d_2^{cr} when the dislocations at $d \leq d_2^{cr}$ remain only at the grain boundaries. The parameters of the critical dimensions of the grains in the microrange are published in [24, 28]. In the following section, the critical grain sizes are examined in a separate study. It is important to stress here again that the relationship (3.11) relates to the microrange of the grain sizes. The following relationship is fulfilled in the mesorange of the grain sizes:

$$\rho = k\varepsilon d^{-1}. \qquad (3.12)$$

where ε is the strain, k is a constant. This relation was introduced by M.F. Ashby [31] and confirmed by H. Conrad [13], A.N. Orlov [33] and the authors of this book [32, 34].

3.2.2. Critical grain sizes

The problem of existence of the critical grain sizes was already discussed in chapter 2. Here, a more generalised pattern of this phenomenon is discussed. The quantitative investigations of the grain structure and the properties of the polycrystals at the microlevel made it possible to define three critical grain sizes. These are the average grain sizes characterised by the large changes in the properties of the polycrystalline aggregate.

The first critical grain size d_1^{cr} is the size at which the sign of the Hall–Petch relation k changes in the following relationship:

$$\sigma_{YS} = \sigma_0 + kd^{-1/2}, \qquad (3.13)$$

where σ_{YS} is the yield limit, σ_0 is the deformation resistance of the single crystal; d is the mean grain size. At $d > d_1^{cr}$ $k > 0$, at $d < d_1^{cr}$ $k < 0$. The value d_1^{cr} for the pure metals Al, Cu, Ni, Fe, Ti is close to 10 nm [35, 36]. The change of the sign of the coefficient k indicates the change of the grain boundary hardening to grain boundary softening. In other words, further grain refining results in a decrease of the yield strength and not increase.

The second critical grain size d_2^{cr} is associated with the formation of dislocation-free grains [15]. The interaction of the grain boundaries with the dislocations becomes so strong that the dislocations are displaced from the body of the grain by the stress fields generated by the grain boundaries, especially the stress fields from the steps at the grain boundaries and triple junctions. The dislocations then move to the grain boundaries. For the pure metals, d_2^{cr} is approximately 100 nm [37, 38]. The dispersion of the grain sizes is close to $\sigma_d =$ 100 nm. The formation of dislocation-free grains hardens the sub-micropolycrystal and causes changes in the deformation mechanism of the latter.

The third critical grain size d_3^{cr} is associated with the change of the role of the parameters of the dislocation structure [24]. If $d > d_3^{cr}$, then the dislocation ensemble contains mostly statistically stored dislocations with the density ρ_s. The number of these dislocations is greater than that of the geometrically necessary dislocations ρ_G ($\rho_s > \rho_G$). The value d_3^{cr} is close to 1000 nm. The passage through this grain size ($d < d_3^{cr}$) changes the origin of a large part of the dislocations, the conditions of screening the stress raisers by the dislocations, and the level of the internal stress fields. In these conditions, the density of the geometrically necessary dislocations becomes greater than the density of the statistically stored dislocations, since $\rho_G > \rho_s$.

In this section, attention is given to the critical grain sizes determined in experiments on pure metals and alloys. Also, attention is given to the problem of internal stresses which are formed by the geometrically necessary dislocations.

The first critical grain size

The Hall–Petch coefficient (H–P) k [39, 40] is a very important characteristic of grain boundary hardening. It determines the increase of the yield strength and the fracture stress with a change of the grain size. The dependence of the H–P coefficient on the grain size has been determined mainly by studies [35, 36]. It is well-known that at a certain small grain size the H–P coefficient k becomes equal to 0. There is no grain boundary hardening in this case. The data for the main pure metals are presented in Fig. 3.7. The graph shows clearly that the conversion of k to 0 takes place at $<d> \approx 10$ nm. It is curious that this number does not depend on the type of metal. It is justified to assume that $<d> = 10$ nm is a fundamental parameter, and the deformation mechanisms of the nanopolycrystals change in the vicinity of this parameter. This is accompanied by a large increase in the importance of the grain boundary deformation processes [15, 41]. These processes include diffusion processes at the grain boundaries, sliding of the lattice

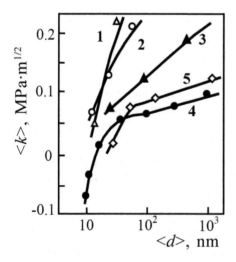

Fig. 3.7. Dependence of the Hall-Petch coefficient k on the grain size for the main metals: 1 – Fe; 2 – Ti; 3 – Ni; 4 – Cu; 5 – Al.

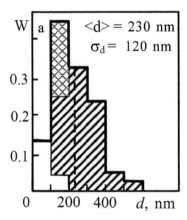

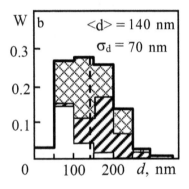

Fig. 3.8. Size distribution of the grains in the sub-micropolycrystalline copper (a) and nickel (b) produced by ECAP (a) and THP (b): ☐ the dislocation-free grains; ☒ the grains with the random distribution of the dislocations or dislocation network; ☒ - grains with dislocation cells or fragments.

and grain boundary dislocations on them, migration of the grain boundaries, etc.

The second critical grain size

The typical size distribution functions of the grains for the sub-micropolycrystalline copper and nickel are presented in Fig. 3.8. The graph shows the dependences for the sub-micropolycrystals of copper produced by the equal channel angular pressing and the sub-micropolycrystals of nickel produced by torsion under hydrostatic pressure. Three types of grains, usually obtained after severe plastic deformation (SPD), are defined [15, 37]. They are characterised by different dislocation structures, namely: 1) dislocation-free grains; 2) the grains with the random dislocation structure; 3) the grains containing a dislocation substructure – cells or fragments (Fig. 3.9). The average sizes of each type of grain are presented in Table

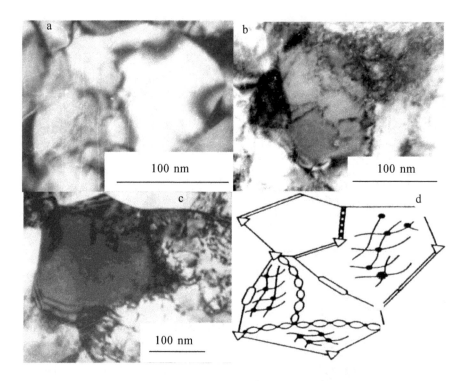

Fig. 3.9. Electron microscopic images (a–e) and the diagram of the three types of grains (d): a – dislocation-free grains; b – the grains with the randomly distributed dislocations or dislocation networks, c – the grains with a dislocation cells or fragments; the diagram (d) shows the nanoparticles distributed inside (●), at the boundaries (⬭) – and at the joints (Δ) of the grains.

3.1. The average grain size of each type of grain increases in the direction from the dislocation-free grains to the grains with the cells and fragments (Table 3.1). The structure of the grains and the distribution of the size of the grains are determined by the methods of preparation of the nanostructured polycrystalline aggregate [42, 43]. At some moment deformation was interrupted and, therefore, the structure contains dislocation-free fine grains and the grains of average sizes with the developed dislocation structure. The largest grains of the nanocrystalline aggregate are found in the cells which gradually transformed during deformation to fragments, i.e., the subgrains, surrounded by the low-angle boundaries. In subsequent deformation the latter transformed to new nanograins. Thus, the size distribution function of the nanograins illustrates the mechanism of their formation during severe plastic deformation. The dislocation-free grains grow as a result of the migration of the grain boundaries.

Table 3.1. Average sizes of different types of grains in micropolycrystalline copper produced by equal channel angular pressing, in nickel produced by torsion under hydrostatic pressure and the average sizes of the fragments in the BCC steel

Material	Average grain size (fragment), nm	Dislocation-free grains (fragments), nm	Grains (fragments) with dislocation chaos	Grains (fragments) with dislocation cells, nm
Cu	230±60	78±33	167±17	316±45
Ni	140±35	84±42	150±26	166±29
BCC steel	260±89	60±13	420±71	230±66

Intragranular deformation leads to the formation in them of initially a chaotic dislocation structure and then a cellular dislocation structure. The cell boundaries then become disoriented, undergo transformations to the low-angle sub-boundaries of the fragments. With further deformation the misorientation at these boundaries increases. They transform to the boundaries of the nanograins.

The interactions of the dislocations with the sources of the internal fields of the stresses and, in particular, with the grain boundaries have a significant effect on the dislocation structure and the mechanisms of deformation of sub-micropolycrystalline aggregates. In the polycrystals the microlevels of the grain boundaries are not only the sources of dislocations but also sinks for them [35]. The latter effect influences the dislocation density and leads to the formation of dislocation-free grains.

Figure 3.10 shows the dependence of the average scalar density of dislocations (ρ), typical of different types of grains, on the average size (d) of the grains of the same type. With a decrease of the size of the ultrafine grains the dislocation density decreases. This takes place firstly with the average dislocation density presented in dependence on the average grain size d (Fig. 3.10a). The dotted lines, extending these relationships, relate to the grain size $d = 16$ nm and 80 nm at a dislocation density of $\rho \approx 0$ (1 and 2 in Fig. 3.10a). Thus, the first critical grain size is found in the vicinity of 50...100 nm. This value for the critical grain size is obtained by averaging with respect to all types of grains. Figure 3.10b shows the dependences $\rho = f(d)$ for the grains with a random dislocation structure (3) and the grains with the fragmented dislocation structure (4). The specimens were produced by THP. Here the dashed lines are longer because the grains

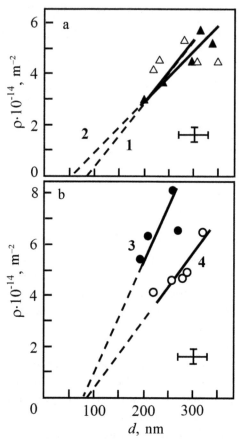

Fig. 3.10. Dependence of the scalar dislocation density ρ on the average grain size d (1 and 2) and on the grain size with the chaotic dislocation structure (3) and the fragmented substructure (4). The dotted line indicates the critical grain sizes: 2 – Ni, ECAP; 1, 3, 4 – Cu, THP.

with these structures do not transform to dislocation-free grains. Nevertheless, the critical grain size can be seen, it is close to 70 nm and 80 nm. The dislocation-free grains are observed when their size is smaller than the second critical size which for pure metals is close to $d_2^{cr} \leq 100$ nm. As mentioned previously, this phenomenon is characterised by the greater dispersion; $\sigma_d \approx 100$ nm. The concept of the critical grain size for the dislocation-free grains was introduced for the first time in [37]. The results presented here confirm the previously published value of d_2^{cr}, close to 100 nm [15]. The same studies give the range of the dimensions of dislocation-free grains. In most cases, the range of the dimensions of the dislocation-free grains extends from the smallest sizes to $d \approx 200$ nm, regardless of the type of material, for both copper and nickel (Fig. 3.7). This indicates the same mechanism of the phenomenon.

The dislocation-free grains play a special role in the formation of the mechanical properties of the sub-microcrystals. Because of the small size, these grains harden the polycrystal. The number of the dislocation-free grains in relation to the total number of the grains varies from 0.10 to 0.40. The volume fraction of these grains doe not exceed 0.05–0.20 because of the small grain size.

3.2.3. Geometrically necessary and statistically stored dislocations, the second and third critical grain sizes. Comparison of the parameters of the micro- and mesolevel

The dislocation structure is characterised by several parameters. In addition to the scalar dislocation density (ρ), an important role is played by the density (ρ_s) of the statistically stored dislocations (SSD), and the density (ρ_G) of the geometrically necessary dislocations (GND). The GND, introduced many years ago by Ashby [3, 31] are now in the centre of attention of investigators [8, 9, 44]. The value ρ_G determines the inhomogeneity of deformation and its gradients [3, 4], the internal stress fields and their screening [45, 46]. This is especially important in investigating the nanopolycrystals [47]. As shown previously, the density of the GND is also the density of excess dislocations. The authors measured ρ_G (or $\rho_\pm$), using the parameter χ:

$$\rho_G = \rho_\pm = \frac{\chi}{b}. \qquad (3.14)$$

The value ρ_s can also be measured using the data for the scalar density of the dislocations ρ and the values for ρ_G, because:

$$\rho = \rho_s + \rho_G. \qquad (3.15)$$

Therefore:

$$\rho_s = \rho - \rho_G. \qquad (3.16)$$

The grain sizes (d) at the nanolevel can be compared with the sizes (d_{fr}) of the dislocation fragments at the mesolevel. Their ranges practically overlap. This may be indicated by comparing Figs. 3.10 and 3.11. Figure 3.11 shows the dependences of ρ, ρ_s, and ρ_G on the size of the fragments in the deformed BCC steel. These dependences are interesting. All three components of the dislocation structure decrease with decreasing fragment size. The extrapolation lines (see the dotted lines in Fig. 3.11) indicate the critical size of the fragments equal to approximately 100 nm. This is a very important result indicating the intensive interaction of dislocations with the fragment boundaries. Undoubtedly, the nature of the phenomenon is the same in both Figs. 3.11 and 3.10. It should be mentioned that three types of fragments are found in the deformed steel: 1) dislocation-free, 2) with the random dislocation structure, and 3) with the cellular dislocation substructure (Fig. 3.12). It is obvious that the structures of the dislocation fragments and the micrograins have many similar features. The dependences of the dislocation density on the size of the fragments and the size of the micrograins also coincide. The critical grain sizes and the sizes of the nanofragments are similar. Obviously, the grain size on the microlevel and also of the fragments determines the identical behaviour of the parameters of the dislocation structure in them.

Since the nanograins in the micropolycrystal and the fragments in the BCC steels are divided into three structural types (Figure 3.12), their sizes should be compared. The following structural types are defined: 1) the small size – clean or dislocation free,

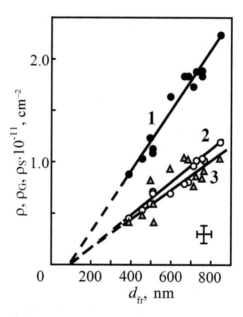

Fig. 3.11. Dependences of ρ (1), ρ_s (2) and ρ_G (3) on the size of the fragments d_{fr} in the deformed BCC steel.

grains or fragments, 2) with a random dislocation structure, 3) with a cellular substructure. Table 3.1 compares the grain sizes in the micropolycrystals of Cu and Ni and the fragments in the BCC steel. The average grain and fragment sizes are comparable. The clean, dislocation free grains and fragments have similar sizes. The ranges of the sizes of the dislocation grains and the fragments overlap. Undoubtedly, the range of the sizes of the grains and the fragments 50...400 nm determines the identical behaviour in both the micro-polycrystals of the pure metals and in the fragmented steel (Table 3.1). The critical sizes of the grains and the fragments are also

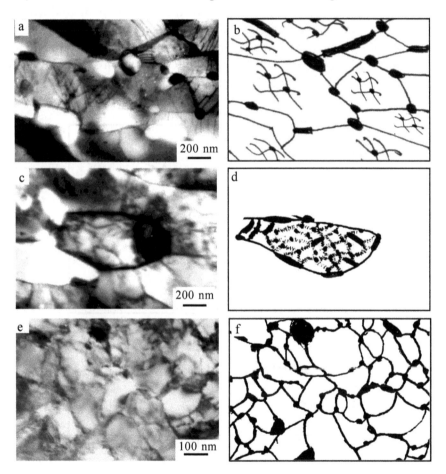

Fig. 3.12. Electron microscopic images of the three types fragments and their diagrams in the deformed BCC steel: a, b) the fragments with a random dislocation structure; c, d) with a cellular substructure in: e, f) dislocation-free fragments.

comparable (compare Figs. 3.10 and 3.11). Therefore, this section presents the efficient comparison of the dimensions and properties of the sub-microcrystalline grains and the nanofragments in the steels.

Here it is necessary to compare the dependence of the dislocation density on the grain size at both the micro- and mesolevel. As an example, the data for the mesolevel are presented in Fig. 3.13 [36]. If the grain size refining on the microlevel leads to a decrease of the dislocation density, then at the mesolevel the effect is opposite: with grain refining the dislocation density increases (compare the data in Figs. 3.10 and 3.11, on the one hand, with the data in Fig. 3.13 on the other hand). The differences in the behaviour of dislocation density in dependence on the grain size result in the difference of the microlevel (or the nanolevel) from the mesolevel. This difference characterises the critical behaviour of the dislocation structure in the polycrystals at the micro- and mesolevel.

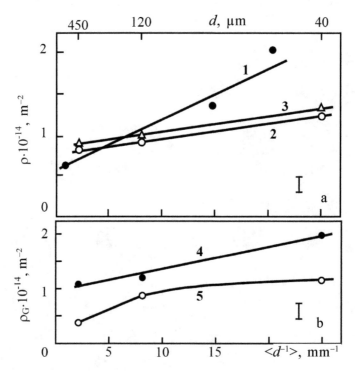

Fig. 3.13. The dependences of the dislocation density on the grain size d of the mesolevel: a – the scalar dislocation density (ρ); b) the density of the geometrically necessary dislocations (ρ_G). The Ni$_3$Fe alloy: 1, 4) tensile loading; 2, 3 and 5) compressive loading; 1, 2 and 4) the alloy with the short-range atomic order; 6, 5) with the long-range atomic order.

The third critical grain size

The third critical grain size (d_3^{cr}) is associated with the geometrically necessary dislocations. The density of these dislocations ρ_G can be considerably smaller than the scalar dislocation density or can be comparable with it. In the first case, the density of the statistically stored dislocation (SSD) provides the main contribution to the dislocation structure. This is characteristic of the polycrystals at the mesolevel. The micro- or nanolevel differs from the mesolevel by the considerable contribution of the value of ρ_G to ρ. The refining of the grains, fragments and cells increases the density of the GNDs. The increase of the density of the nanoparticles also increases the density of the GNDs. It is justified to assume that the relatively high value of ρ_G is the characteristic difference of the nanopolycrystals from the polycrystals on the mesolevel. The third critical grain size corresponds to the equality of both components of the structure $\rho_S = \rho_G$. When $\rho_S > \rho_G$, these are the polycrystals on the mesolevel. When $\rho_S < \rho_G$ these are the polycrystals on the micro- or nanolevel. Figure 3.14 shows the appropriate data which can be used to compare the values of ρ, ρ_S and ρ_G in a wide range of the sizes of the grains and fragments. Figure 3.14 shows that the third critical grain size is in the range 5...10 μm. If $d < d_3^{cr}$, the controlling role in the dislocation structure is played by the GNDs. The formation of the main and accommodation slip systems in the grains [49–51] is prevented and the microlevel realised.

The transition from the mesolevel to the microlevel in the entire range of the size of the grains and the fragments is presented using the GND in Fig. 3.15. If at the mesolevel ρ_G is 0.1...0.2 of the value of ρ, then on approaching the grain size of 200...300 nm the dislocation structure is formed completely ($\rho \approx \rho_G$). The gradient dislocation structure forms as a result of the stress fields from the disclinations, situated at the grain boundaries and triple junctions of the grains,

3.3. Dependence of the scalar density of the dislocations on the size of the fragments with the network dislocation substructure in a martensitic steel

The typical pattern of the fragments with the network dislocation substructure of the deformed steel is presented in Fig. 3.16. The

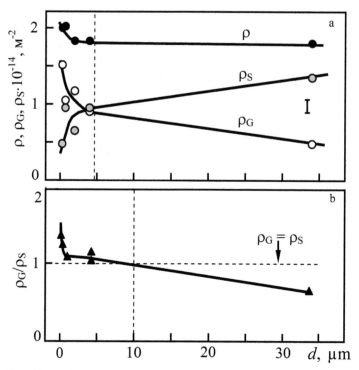

Fig. 3.14. Effect of the size (d) of the fragments (a) and the grains (b) on the magnitude of the different components of the dislocation structure, ρ_G for the deformed BCC steel. The vertical dotted line indicates the third critical size of the dislocation fragments (a) and grains (b).

distinctive relationships between the dislocation density and the size of the fragments were determined for the fragments with the network dislocation substructure in the martensitic steel. The scalar density of the dislocations and the size of the fragments d_{fr} in the steel are linked by the relationship:

$$\rho = C' d_{fr} \tag{3.17}$$

where C' is a constant. The dependence of the scalar density of the dislocations on the size of the fragments of the tempered deformed martensitic steel is governed by the same relationships as the dependence of the scalar density of dislocations on the grain size of the ultrafine-grained structure of the pure metals Cu and Ni [24, 27, 28]. The similarity in the behaviour of the dependences $\rho(d)$ and $\rho(d_{fr})$ in Figs. 3.10, 3.11 and 3.17 indicates the presence of the fundamental relationships linking the scalar density of the

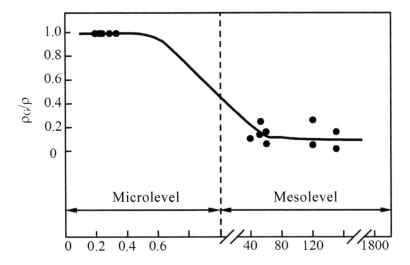

Fig. 3.15. The relationship of the density of the geometrically necessary dislocations ρ_G with the scalar dislocation density ρ in a wide grain size range.

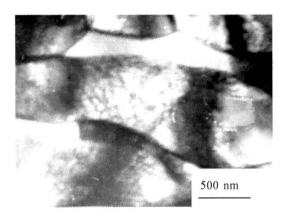

Fig. 3.16. The network dislocation structure inside fragments of the deformed steel (transmission electron microscopy).

dislocations with the size of the grains of fragments in which the dislocations are stored with deformation. The typical dependence $\rho = f(d_{fr})$ is presented in Fig. 3.17. The figure indicates that the relationship (3.17) is strictly fulfilled. At the same time, Fig. 3.17 demonstrates the critical size of the fragments at which the scalar density of the dislocations becomes equal to 0, $\rho = 0$. This critical

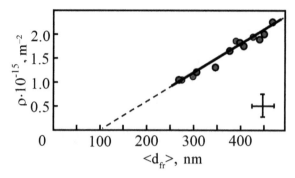

Fig. 3.17. Dependence of the scalar dislocation density (ρ) on the average fragments size ($<d_{fr}>$) with the network dislocation structure.

size is close to the size $d_{fr}^{cr} = 100$ nm. At the same time, it has been established that the second critical size of the grain in the micro-region is equal to the critical size of the fragments in the dislocation structure of the martensitic steel formed during deformation. It is believed that the equality of the critical sizes of the micrograins and the dislocation fragments is completely determined by the similarity of the mechanisms of interaction of the sliding dislocations with the boundaries of the micrograins and fragments.

3.4. Dependence of dislocation density on the size of fragments with the cellular dislocation substructure in the martensitic steel

In transition to the cellular dislocation substructure the pattern of the relationships of the storage of dislocations in the steel becomes more complicated. The dislocation cellular substructure, formed inside the fragments of the deformed martensitic steel, is shown in Fig. 3.18. In this case, a decrease of the size of the fragments does not decrease the dislocation density which in fact increases. Undoubtedly, this is associated with the fact that the sliding dislocations interact with the walls of the cells situated inside the fragments, in a different manner in comparison with the dislocations of the network dislocation substructure. The barrier inhibition [52] by the walls of the cells is a stronger factor, and with a decrease of the size of the fragments, containing the cellular substructure, the scalar density of the dislocations increases. The appropriate data are presented in Fig. 3.19. It may be seen that the following relationship is satisfied

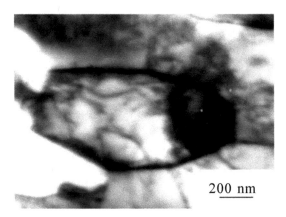

Fig. 3.18. The cellular dislocation substructure insider fragments of the deformed martensitic steel (transmission electron microscopy).

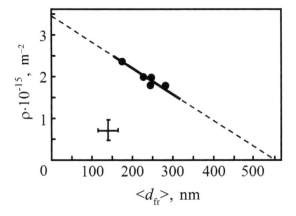

Fig. 3.19. Dependence of the scalar dislocation density (ρ) on the average size of the fragments ($<d_{fr}>$) with the cellular dislocation structure.

$$\rho = A'D_{cell}^{-1}, \qquad (3.19)$$

where A' is a constant. Comparison of Figs. 3.19 and 3.20 shows that the dependence of the dislocation density on the cell size is stronger than the dependence on the size of the fragments ($A' > A$). This means that the main effect is the barrier inhibition of the sliding dislocations by the walls of the cells. The extrapolation of the dependence, presented in Fig. 3.20, indicates the limiting value $\rho > 5 \cdot 10^{11}$ cm^{-2}. This high scalar density of dislocations, close to $\rho \approx 10^{12}$ cm^{-2}, indicates that the inhibition of the dislocations by

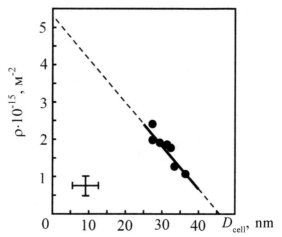

Fig. 3.20. Dependence of the scalar dislocation density (ρ) on the cell size (D_{cell}) in the deformed martensitic steel.

the walls of the dislocation cells should be related to the maximum inhibition of the sliding dislocations in the dislocation structure.

In this section, using the direct experimental data, we present the analytical dependence of the scalar density of the dislocation on the size of the fragments and cells in the deformed martensitic steel. The results show that the type of dislocation structure in the fragments of the steel has a controlling effect on the dependence of the scalar density of the dislocations on the size of the dislocations. If there is a network dislocation substructure insider fragments, the scalar density of the dislocations decreases with a decrease of the size of the fragments. On the other hand, if the dislocation structure in the fragments cellular, then a decrease of the size of the fragments increases the scalar dislocation density. The dislocation density also rapidly increases with a decrease of the cell size. This behaviour is determined by different mechanisms of inhibition of sliding dislocations in the network and cellular dislocation substructures. The determine the relationships between the different parameters of the dislocation structure are fundamental for further development of the physics of dislocation hardening of solids [53, 54]. The dislocations paradigm [54] of the substructural hardening obtained in these relationships is, firstly, the direct confirmation of the importance of the physics of dislocations and work hardening of the materials and, secondly, describes the fundamental role of the substructural formations in the dislocation concept of the physics of work hardening.

3.5. Effect of the size of the fragments of grains and on the density of defects in metallic materials

The deformation of the polycrystalline aggregates is usually accompanied by the formation of the dislocation structure. The scalar dislocation density (ρ) in the polycrystals depends on the average grain size (d). This dependence forms as a result of the number of factors, in particular, the path length of the dislocations in the grains which is restricted by the grain size, the density of dislocation sources at the grain boundaries, the effect of the grain size on the formation of dislocation structures and the rate of annihilation of the dislocations [32, 55, 56]. Another important factor is the interaction of the dislocations with the grain boundaries and, consequently, the transfer of part of the lattice dislocations to the grain boundaries with subsequent transformation of the structure up to the complete dissolution at the boundaries [57].

In addition to this, the scalar density also depends on the type of grains. It was established in [15] that there are three types of grains in the ultrafine-grained polycrystals: 1) the dislocation-free grains; 2) the grains with the chaotic dislocation substructure, and 3) the grains with the cellular and fragmented substructure. The size of the grains increases in the order of numeration of these substructures.

The nature of the functional dependences of the scalar density of the dislocations on the average grain size differs for the polycrystals on the meso- and microlevels. In the case of the polycrystals on the mesolevel the grain refining results in an increase of the scalar dislocation density [48, 50], and for the polycrystals on the microlevel the scalar dislocation density decreases with a decrease of the grain size up to the formation of completely dislocation-free grains [24, 26, 59, 60, 61].

The experimental data indicate that the dislocation density depends not only on the grain size but also the size of the subgrains, fragments, cells and other structural formations [62]. In the martensitic steels they also include laths, plates and twins. The authors investigated the relationships governing storage of dislocations during deformation independence on the size of the micrograins and fragments in different metallic materials.

Several materials were investigated. Firstly, it was the ultrafine-grained copper, prepared by two methods: 1) equal channel angular pressing (ECAP), and 2) torsion under quasi-hydrostatic pressure (THP). The specimens of the ultrafine-grained copper, prepared by the ECAP method and having the size of 4×4×6 mm³, were deformed by compression in an Instron machine at room temperature in the strain range ε = 0...83%. In the THP method the specimen was cylindrical, 0.2 mm long, diameter 10 mm, and was subjected to plastic deformation at a temperature of T = 293 K, pressure P = 5 GPa. Logarithmic strain was ε = 5. Secondly, the ultrafine-grained nickel, prepared by ECAP. The specimens of the ultrafine-grained Ni were investigated in two conditions: 1) immediately after preparation, and 2) after short-term annealing at T = 398 K, used to decrease the amplitude of internal stresses. In addition to the ultrafine-grained copper and nickel, the 0.34C–1Cr–3Ni–1Mo–1V–Fe tempered martensitic steel was also investigated. The specimens of the steel after quenching and subsequent high-temperature tempering were tensile deformed in Instron equipment at room temperature in the strain rate range ε = 0...90%.

The main method of investigating the structure of these materials was transmission diffraction electron microscopy. The structure of the materials was investigated on thin foils prepared by electrolytic polishing in special conditions producing large areas for examination. The foils were examined in an EM-125 K electron microscope using a goniometric attachment. The following parameters were measured: for the ultrafine-grained copper and nickel – the average grain size and the dimensions of grains of different types with different dislocation structures; in steels, the dimensions of the dislocation fragments were also measured. In all types of grains of the ultrafine-grained copper and nickel and in the steel fragments measurements were taken of the scalar dislocation density. Both the average dislocation density in the polycrystal as a whole and the dislocation density in each type of grain separately were investigated. The statistical processing of the results was carried out in continuous sections of the specimens with the area of 80 μm², containing 500–1000 grains or fragments. The dimensions of the grains, fragments and scalar density of dislocations were taken by the secant method [63, 64]. Using the parameters of the bending extinction contours, observed on the electron microscopic images of the structure, measurements were taken to determine the curvature–torsion (χ) of the crystal lattice [7, 10]. This relationship should be mentioned again:

$$\chi = \frac{\partial \varphi}{\partial \ell}. \qquad (3.20)$$

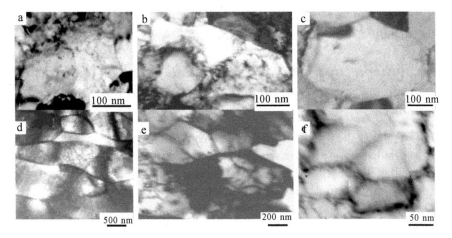

Fig. 3.21. Electron microscopic images of different types of grains (a – with the chaotic distribution of the dislocations, b – fragmented grains, c – dislocation-free grains) in the ultrafine-grained polycrystals of copper and fragments in the deformed tempered martensitic steel (d – the fragments with the network substructure, e – with the cellular substructure, f – dislocation-free substructure).

Table 3.2. Classification of the size of the grains and fragments

Scale level	Metal polycrystals		Fragments in steel	
	Type of polycrystal	Average grain size d	Structural element	Average size of element d
Mesolevel	Coarse-grained (macro) polycrystal	0.1–10 mm		
	Conventional (meso) polycrystal	10–100 μm		
	Fine-grained polycrystal	1–10 μm	Packet	4 × 6 μm
			Plate	1...2.5 μm
Microlevel	Ultrafine-grained (UFG) polycrystals	0.2–1 μm	Fragment with network DSS	0.2 × 1 μm
			Fragment with cellular DSS	50 × 400 nm
	Sub-micropolycrystals	50–200 nm	Dislocation-free fragment	50...70 nm
	Nanopolycrystals	3–50 nm	Cell	25...35 nm

Here φ is the angle of inclination of the crystallographic plane, ℓ is the distance on the plane.

3.5.1. Similarity of the dimensional relationships in ultrafine-grained polycrystals of metals and steels with a fragmented structure

Figure 3.21 shows the typical electron microscopic images of the closed substructural formations in the investigated materials in which the scalar dislocation density was measured. The classification of the polycrystals on the basis of the grain size was described previously. This classification also includes the dimensions of substructural formations in the martensitic steel. The data are presented in Table 3.2.

It should be noted that the dislocation structure of the fragments in the steel is similar to the dislocation structure of the grains in ultrafine-grained polycrystals. The steels contain dislocation-free fragments, fragments with a chaotic dislocation structure and fragments with a cellular substructure.

The grain size of the ultrafine-grained and nanopolycrystalline materials is in the range 5...500 nm [15, 65]. This size range can be compared with the dimensions of dislocation cells, fragments and the width of the laths in the martensitic steel [62]. The size of the latter is in the range 30–400 nm (Table 3.2).

The measurements show that the dependences of the scalar density of dislocations on the size of the different structural formations in different materials have much in common. The size of the fragments influences the scalar density of the dislocations like the size of the grains and subgrains. The appropriate data are presented in Fig. 3.22. Firstly, the figure shows the dependences of the scalar dislocation density on the average grain size for the ultrafine-grained polycrystals of copper and nickel. Secondly, there are also the dependences of the scalar dislocation density for the fragments with the network substructure in the deformed tempered martensitic steel. In all cases, the scalar dislocation density is proportional to the size of the grains, i.e., the following relationship applies (presented again)

$$\rho = C \cdot d, \tag{3.21}$$

where C is a constant.

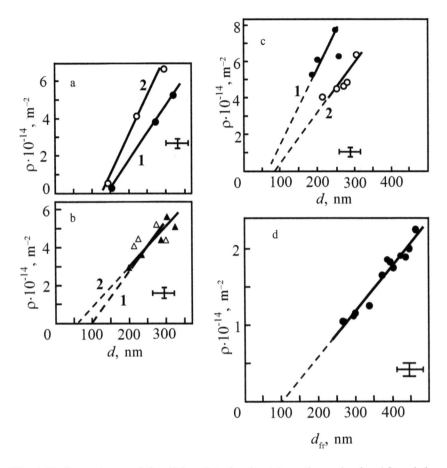

Fig. 3.22. Dependence of the dislocation density (ρ) on the grain size (d) and the size of the fragments (d_{fr}) in different materials: a – for the average grain sizes of the ultrafine-grained polycrystals of nickel (1 – freshly prepared condition after, ECAP, 2 – the condition after subsequent annealing); b – for the average grain sizes of the ultrafine-grained polycrystals of copper, prepared by THP (1) and ECAP (2) methods; c – for the grain sizes of the ultrafine-grained polycrystals of copper with the chaotic dislocation structure (1) and for the grains with a fragmented substructure (2); d – for the fragments with the network substructure in the deformed tempered martensitic steel.

In particular, it should also be mentioned that this dependence does not correspond to the relationship proposed by Ashby [3.31] and Conrad [13]

$$\rho = B \cdot d^{-1}, \tag{3.22}$$

where B is a constant. It should be noted that the relationship (3.21) was in fact proposed by the authors of [3, 13] for the grains on the mesolevel. For the grains on the mesolevel this relationship is in fact fulfilled [10, 58, 65, 67]. For the grains on the microlevel the relationship (3.22) is not fulfilled, and the relationship (3.21) holds.

It is important to note that the dependence of the dislocation density ρ on the size of the fragments in the steel is similar to the dependence of ρ on the size of micrograins in the ultrafine-grained metals. The dependence of the scalar dislocation density on the size of the fragments in the steel is presented in Fig. 3.22. It may be seen that the value of ρ decreases with a decrease of the fragment size, i.e., the dislocation density in the fragments behaves in the same manner as in the ultrafine-grained and nanopolycrystals. As indicated by Fig. 3.22, the relationship (3.20) is fulfilled for the ultrafine grained and nanopolycrystals of the pure metals and fragments. Thus, the experimental results show that different relationships may be fulfilled between the dislocation density in the grain size, either (3.21) or (3.22). It is evident that it is necessary to find the dimensional coefficients for these relationships.

The results indicate that the dimensional effect is practically the same in both the ultrafine-grained and nanopolycrystals of metals, and fragments of the steel. It appears that in the metallic materials the controlling role in the defective structure is played not only by the types of closed structural formations but also their size. In particular, these dimensions basically determine the scalar dislocation density stored in the process of plastic deformation. The experimental quantitative dependences of the dislocation density on the size of the closed substructural formations lead to the conclusion on the similarity of the dimensional relationships in the ultrafine- grained polycrystals and fragments of the martensitic steel. Figure 3.22 shows clearly the critical size of the grains and fragments at which they become dislocation-free. For this purpose, the dotted lines in Fig. 3.22 extend the appropriate relationships to the value $\rho = 0$. The critical sizes of the grains and fragments coincide. Their values are close to 100 nm.

This section presents the quantitative investigations of the effect of the size of the grains and fragments in some materials on the scalar density of the dislocations. The results show that at the grain and fragment sizes smaller than 1 µm the density of dislocations decreases with a decrease of the grain size both in the ultrafine-grained polycrystals and in the dislocation fragmented substructure of the steels. The critical size of the closed substructural formations

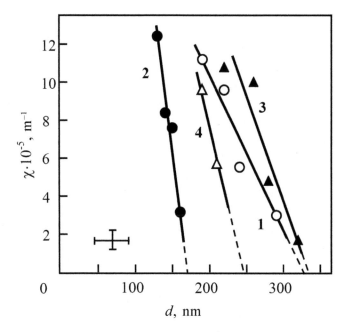

Fig. 3.23. Dependence of the curvature–torsion of the crystal lattice in grains with different dislocation substructures on their average size d: 1 – the average value of χ for all types of grains; 2 – in the dislocation-free grains; 3 – in the fragmented grains; 4 – in the grains with a random dislocation structure. Copper was prepared by the THP method.

(dislocation-free grains in dislocation fragments) was determined, $d_{cr} \approx 100$ nm.

3.5.2. Dependence of the density of partial disclinations on the grain size

The size of the grains in dislocation fragments influences the dislocation density. This was investigated in detail in the previous sections. However, this is not the only effect of the dimensions. The size of the grains and dimensions also influence the disclination density, the curvature–torsion of the crystal lattice and the internal stresses. This problem will be investigated further in the section. The size of the grains and fragments also influences the density of point defects, the free and constricted volumes, the density of the pores and nanoinclusions. The latter problem was investigated in detail in a review by R.A. Andrievskii, concerned with the radiation resistance of nanomaterials [68].

The scalar dislocation density decreases with the decrease of the grain size of the microlevel. This is accompanied by the increase of the curvature–torsion of the crystal lattice (Fig. 3.23), the density of geometrically necessary dislocations and the internal stresses [69]. As this is accompanied by a decrease of the dependence of the scalar dislocation density, it is necessary to answer the question what is the reason for the increase of the curvature–torsion of the crystal lattice and the internal stresses. Since the scalar dislocation density decreases, this phenomenon may be caused by the increase of the density of partial disclinations. The decrease of the density of dislocations does not indicate the formation of defect-free grains, fragments, etc. In many cases, especially in nanomaterials, the decrease of the scalar density of the dislocations is compensated by the increase of the density of disclinations. In the ultrafine-grained polycrystals, they are preferentially distributed in the triple and quaternary junctions of the grains [70, 71].

Figure 3.24 shows the dependence of the density of partial disclinations in relation to the size of the grains of ultrafine-grained copper, prepared by the THP method. The density of the partial disclinations was determined from the density of the triple junctions containing these disclinations. The typical size of the triple junctions of the grains, containing the triple disclinations in the ultrafine-grained copper polycrystals is shown in Fig. 3.25. The bending extinction contours spreading from the triple junctions are clearly visible. This indicates that the source of internal stresses are the triple junctions and, in particular, the triple disclinations, distributed at these junctions. Figure 3.24 shows for comparison the dependence of the scalar density of dislocations on the grain size. It is clearly seen that the behaviour of the density of disclinations in the ultrafine-grained polycrystals is opposite to the behaviour of the dislocation density. In the grain size range 100...400 nm the scalar density of the dislocations decreases and the density of disclinations increases. Gradually, with a decrease of the grain size due to the decrease of the scalar density of dislocations they cease to be the main defects of the crystal lattice. The main defects are then partial disclinations, situated at the junctions of the grains. With a decrease of the grain size this effect increases the internal stresses, the curvature–torsion of the crystal lattice, the defectiveness of the triple junctions, etc. [43, 70, 71]. This takes place in the grain size range 400...200 nm. After $d \leq 100$ nm the grains become dislocation-free and the density of partial disclinations continues to increase.

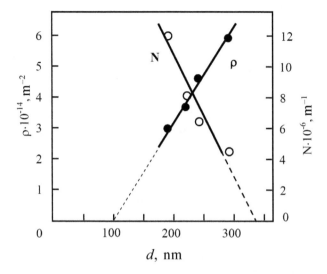

Fig. 3.24. Dependence of the scalar dislocation density (ρ) and the linear density of partial dislocations (N) at the junctions of the grains on the average grain size (d), the copper was prepared by the THP method.

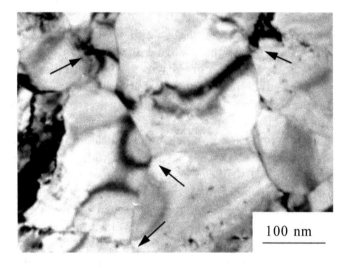

Fig. 3.25. Extinction bending contours, emitted from the triple junctions of the grains (indicated by the arrows), indicating the presence of disclinations in the ultrafine-grained polycrystals of copper, produced by the THP method.

A decrease of the grain size increases the disclination density at the junctions of the grains and, correspondingly, the amplitude of the curvature–torsion (χ) of the crystal lattice. The quantitative relationships governing the change of the parameters of the defective structures are similar in the ultrafine-grained polycrystals and in the fragmented substructure. This shows the important role of the dimensional effect in the formation of the dislocation–disclination structure. The role of the joint disclinations in the polycrystals at the nanolevel will be investigated.

In accordance with the resultant electron microscopic images of the structure of nanocrystalline copper after THP (see section 2.5), the following classification of the triple junctions has been proposed [70]: 1) the junctions with dislocations; 2) the junctions with stressed nanoparticles; 3) 'clean' junctions (without particles and disclinations); 4) the junctions with non-stressed particles.

In the junction with the dislocations there is a discrepancy of the crystallographic orientation of the contacting grains. This discrepancy is compensated by a partial disclination. Usually the electron microscopic images show extinction contours emerging from these junctions or surrounding them. The typical patterns together with the schemes are shown in Fig. 2.12. Here, the dislocation-free extinction contour passes through the dislocation-free grain in which that is no screening of the field of the junction. The dislocation contour passes through the grain with a dislocation structure which partially screens the field of the junctions.

The junctions, containing disclinations or stressed particles, are one of the most important sources of internal stresses. Only the 'clean 'junctions with the compensated disorientation do not produce internal stresses. In electron microscopic studies the stressed triple junctions in contrast to the non-stressed junctions are accompanied by the formation of bending extinction contours. In the investigated nanocrystalline copper increasing strain results in a decrease of the fraction of the non-stressed junctions, and the fraction of the stressed junctions increases. At the distance $X = 7 \cdot 10^{-3}$ m from the torsional centre of the specimen the fraction of the stressed junctions equals 0.75, i.e., 3/4 of all the junctions. Increase of strain results in a decrease of the size of all types of grains.

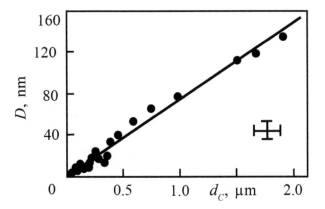

Fig. 3.26. Dependence of the size of nanoparticles (D) and the distance between the sub-boundaries (d_C) at which they are located in the BCC 0.34C–1Cr–3Ni–1Mo–1V–Fe deformed steel.

3.5.3. Particles of second phases, dislocations and boundaries of grains and fragments

The particles of the second phases in the materials nucleate, grow and evolve in close interaction with the defective structure. The different aspects of this process were investigated previously by the authors of this book in [37, 72]. In the previous sections, it was shown that the dislocation structure in the grains usually forms in the constricted conditions for each grain separately. This results in the characteristic dependence of the dislocation density on the grain size. It appears that the formations of the second phase particles is governed in many cases by the same relationships. The size of the particles depends on the size of the grains of fragments and the particles are deformed in the body or at the boundaries of these grains or fragments. This was indicated for the first time in early studies by different authors [37, 73–75]. Figure 3.26 shows the dependence for a BCC steel with the size D of the nanoparticles on the size of the fragments and grains (or the distances d_C between the sub-boundaries), distributed at the boundaries. It may be seen that in the particle size range 10...150 nm the linear dependence $D = f(d_C)$ is valid. This indicates that the particles form from the material of impurity atoms, distributed in each fragment or grain. The dependence $D = f(d_C)$ in Fig. 3.26 is expressed by the equation:

$$D = Bd_C \qquad (3.23)$$

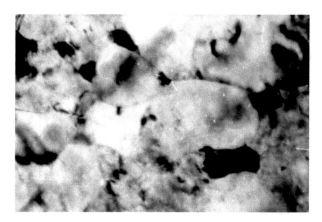

Fig. 3.27. Example of particles of the second phases in the junctions and the grain boundaries in sub-microcrystalline copper: the white arrows indicate the particles at the junctions, the black arrows the particles of the grain boundaries.

where B is a constant quantity. For the investigated steel $B = 0.07$. The relationship (3.23) is one of the many relationships between the size of grains of fragments (d_{fr}) and the size of the particles which stabilise them (D) and are distributed at the boundaries. Other possible relationships will be discussed in section 3.5.7.

3.5.4. Plastic deformation and nanoparticles of second phases in microcrystalline metals

In plastic deformation, the sliding dislocations, the migrating boundaries and sub-boundaries accelerate the processes of redistribution of different impurities in the volume of the material. This relates in particular to the low-solubility substitutional impurities and interstitial elements. During plastic deformation the atoms of these impurities are trapped by defects, in particular by the sliding dislocations and are displaced preferentially to the sub-boundaries, high-angle boundaries and the junctions where they form the second phase particles. Severe plastic deformation results in the rapid displacement in the volume of the material of the substitutional and interstitial impurities. As mentioned previously, these impurities may be present in the initial material or be trapped from the surrounding atmosphere during severe plastic deformation. Detailed electron microscopic diffraction studies revealed nanosized particles of the second phases in the micropolycrystals. Figure 3.27 shows the example of the electron microscopic image of the particles of second phases at the boundaries of grains of microcrystalline copper. Identification of the observed phases showed that the microcrystalline copper contains quite different phases: Cu_2Sn (Sb), Cu_3N, Cu_3O and CuO [75]. These phases are both equilibrium and non-equilibrium. The first phase forms from the impurities present initially in copper, the others – from the nitrogen and oxygen captured from the atmosphere during deformation. Ultrafine-grained nickel contained particles of Ni_4N, Ni_3C, NiO and Ni_2O_3. The mean sizes of these particles were: $Cu_3Sn(Sb)$ – approximately 50 nm, Cu_2N – ~10...40 nm, particles of copper and nickel oxide and also Ni_4N and Ni_3C – 2...8 nm. It may be seen that in the developed plastic deformation conditions the materials contains both stable and low-stability phases. The latter include Cu_3N, Ni_3N and Ni_3C. The phase particles are localised at defects, mostly at the grain boundaries and their junctions. It should be noted that the methods of preparation of the microcrystalline materials influences the volume fraction of the resultant particles: this fraction is greater in THP than in ECAP.

3.5.5. Fragmented dislocation substructure in martensitic steels and second phase microparticles

The mechanism of formation of the second phases is associated with the strong interaction of the defects with impurities and anomalous mass transfer of the impurities in the conditions of the high density of mobile defects. As a result of severe deformation, they contain a fragmented dislocation substructure. The fragments can be both isotropic (equiaxed) and anisotropic (stretched). The anisotropic fragments are primary and contain the dislocation structure (network or cellular). Isotropic fragments form only during deformation and are free from dislocations. Carbide phase particles are found inside and at the boundaries of the anisotropic fragments. In the isotropic fragments the carbides are distributed only at the boundaries and junctions of the fragments. Here, attention is given to the structure of 0.34C–1Cr–1Mn–1Si–Fe and 0.1C–5Cr–1Mo–Fe steels, investigated by the authors of this book in [75], after developed deformation by multipass hot rolling. Developed deformation of the steel resulted in the formation of a fragmented dislocation substructure stabilised by the particles of secondary carbides (special carbides M_2C and M_6C) localised preferentially at the boundaries and in junctions of the fragments. This is indicated by the micrographs shown in Fig. 3.28. The appearance of the secondary nanocarbides indicates that plastic

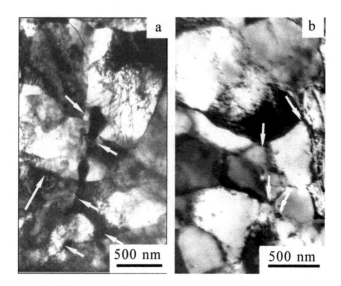

Fig. 3.28. Fragmented substructure in 0.34C–1Cr–1Mn–1Si–Fe (a) and 0.1C–5Cr–1Mo–Fe (b) steels. The arrows indicate the carbides distributed at the fragment boundaries.

deformation was accompanied by structural and phase transformations leading to the formation of a fragmented substructure decorated with carbide particles. The carbon atoms for the formation of secondary carbides during plastic deformation leave the defects and retained austenite. Previously the large part of the carbon atoms was already situated at the sub-boundaries.

In the fragmented substructure the carbide phase is gradually displaced to the boundaries. The maximum stability of the substructure is obtained when all the carbides are distributed at the fragment boundaries. The size of the carbides, situated inside the fragments, is close to 10 nm and is almost independent of the size of the carbides situated at the boundaries. On the other hand, the particles at the grain boundaries are closely linked in their dimensions both with the grains and with the fragments (Fig. 3.26). It may be assumed that the value 10 nm is the critical size of the nanoparticles not trapped by the boundaries of the fragments.

3.5.6. Mechanisms of formation of second phase particles at the boundaries of elements of the microstructure

The formation of second phase particles at the boundaries of different type and their joints in thermomechanical treatment of metallic materials is determined in most cases by three mechanisms. The first mechanism is the diffusion displacement of the particles behind the migrating boundary. During this displacement the second phase particles may grow as a result of the atoms of alloying elements or impurities collected by the migrating boundaries during their displacement. The second mechanism is the collection of the atoms of alloying elements or impurities by the sliding dislocations, displacement of the elements or impurities by the dislocations to the boundaries of the grains or sub-boundaries where new particles form subsequently by diffusion. The third mechanism is the bulk diffusion of the alloying elements which transfers in the final stage to grain boundary diffusion. If the particles form during tempering at high temperatures, the controlling contribution is provided by bulk diffusion. When the second phase particles form during low-temperature deformation and subsequent rearrangement of the substructure, the controlling contribution is provided by dislocation mass transfer. The role of the migrating boundaries and diffusion at the boundaries is always important, especially at intermediate temperatures. All the mechanisms operate during the formation of the substructure, and their relative contribution depends on the temperature and the rate of the deformation processes. The dimensions of the resultant second phase particles are basically determined by the number of the atoms of the alloying elements and impurities, trapped by the defects, i.e., by dislocations or sub-boundaries sweeping during their displacement the area of volume of the micrograins and fragments.

3.5.7. Stabilisation of the structure of microcrystals by second phase particles

The stability of the microcrystals is a very important problem because of the large number of possible applications. It is important to know the laws of formation of structures with the ultrafine size of the grains and fragments whose thermodynamic stability is ensured by the second phase particles. This problem has been known for a long time [37, 72]. The attempts to solve the problem have been based on the analysis of the experimental data and theoretical studies of the problem of the interaction of the boundary and the particle inhibiting the boundary [76–80]). Different variants of the formalism of the theory, very similar to each other, were developed previously by Ziener, Gladman, Hillert, Mader, Hornbogen. They have been generalised and analysed in detail in [73, 81–84]. It was shown that the interaction of a particle with a radius r results in the inhibition force F of the boundary bent along an arc with a radius L, in the following form

$$F = A(\Theta)\pi\gamma'\left(\frac{1}{L} - \frac{1}{r}\right),$$ (3.24)

where γ' is the energy of the interfacial surface; $A(\Theta) \approx 1$ is the angular multiplier. Using the equation (3.23) the authors of [73, 74] show that if the grain structure is stabilised by the second phase particles, the following relationships should be fulfilled between the size of the grains (d)) and particles (D) and the volume fraction (δ) of the particles:

$$d = A_1 \frac{D}{\delta}$$ (3.25)

or a similar relationship

$$d = A_2 \frac{D}{\sqrt{\delta}}.$$ (3.26)

In the relationships (3.25) and (3.26) A_1 and A_2 are constant quantities.

As a result of the investigations of sub-micropolycrystalline materials, the equations (3.25) and (3.26) have been recently subjected to experimental verification [73]. The efficiency of these relationships was confirmed using aluminium alloys with the microcrystalline grain size, hardened by zirconium and niobium oxides in [73].

The secondary phases are produced specially in the two-phase microcrystalline aggregates. In martensitic steels, severe deformation at elevated temperatures results in the two-phase fragmented substructure with the ultrafine size of fragments [85]. The alloying elements are redistributed during deformation in such a manner that the boundaries of the fragments, especially at the junctions of the boundaries, contain microparticles of the secondary phases, which restrict the growth of subgrains.

Measurements of the parameters of the grain and fragmented structures and the secondary phases in the Cu and Ni microcrystals were used to verify the relationships (3.25) and (3.26). Figure 3.29 shows the examples of analysis of these relationships for copper,

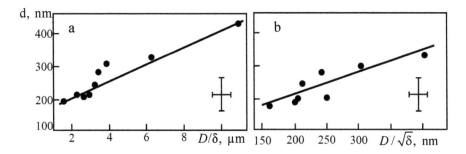

Fig. 3.29. Dependence of the size of the subgrains d for the sub-micropolycrystalline copper on the particle size D: Cu_3Sn (a) and Cu_3N (b) and their volume fraction δ.

produced by the THP method. Analysis of the results shows that, firstly, the relationships (3.25) and (3.26) are satisfied and, secondly, that the phases produced by severe plastic deformation stabilise the structure of microcrystals. The phase particles distributed at the boundaries and the junctions of the grains prevent the displacement of the boundaries. These results are in agreement with the data presented in [73] where the second phase particles were especially added to stabilise the structure of the microcrystals. In this case, the relationships (3.25) and (3.26) between d, D and δ are also satisfied.

The data for steels were quantitatively processed and plotted on the same dependence of the relationship of different parameters of the fragmented substructure. These dependences are presented in Fig. 3.30. In particular, it should be noted that the data for different steels are located on the same relationships. It may be seen that the size of the fragments d_{fr} decreases with increasing volume δ of the second phase (Fig. 3.30). The relationship $d_{fr} = f(\delta)$ in Fig. 3.30 can be described by both a curve and two fragments of the straight lines and the inflection point is located in the centre of the graph. Equation (3.25) is verified in Fig. 3.30. It is fulfilled using two straight branches. The inflection point is situated at $d_{fr} \approx 400$ nm. Similar dependences with the inflection point were found in [73]. It was attempted to straighten the dependences presented in Fig. 3.30a. The appropriate data are presented in Figs. 3.30c, d. It appears that the sizes of the fragments and carbides are linked by a linear relationship $d_{fr} = f(D)$. As regards the relationship of the size of the fragments in the volume fraction of the carbide phase, it is linearised by the dependence

$$d_{fr}^2 = f(1 - \delta). \tag{3.27}$$

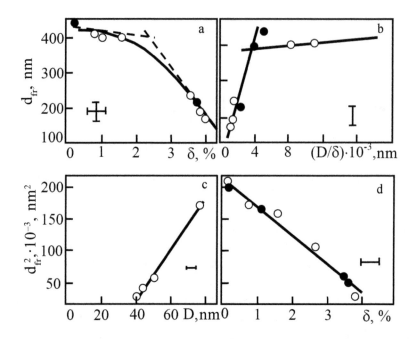

Fig. 3.30. Relationship of the size of the fragments d_{fr}, the dimensions of the special carbides situated at the boundaries, D, and their volume fraction δ: ● – 0.1C–5Cr–1Mo–Fe steel, O – 0.3C–1Cr–1Mn–1Si–Fe.

The general tendency of the stabilisation of the fine-grained substructure with nanoparticles and the fragmented carbide substructure is expressed by the equations (3.24)–(3.27). It may be seen that the physical fundamentals of stabilisation of the nanostructures have been developed. In future, it is necessary to construct a general thermodynamic theory of both the boundaries and the boundaries with the particles.

3.6. The role of geometrically necessary dislocations in the formation of deformation substructures

As already mentioned, the development of the dislocation science has resulted in the sub-division of the value ρ into components of different physical meaning. The scalar dislocation density can be divided to statistically stored dislocations ρ_S (SSD) and the geometrically necessary dislocations (GND) ρ_G.

The GNDs form during deformation in polycrystalline aggregates, in materials with strain twins, in dispersion-hardened materials and in other cases of functioning of strong barriers to dislocation sliding. Other geometrically necessary objects are also studied in addition to the GND. They include the geometrically necessary boundaries (GNB), the geometrically necessary twins (GNT) and other defects. The sub-boundaries and boundaries of the deformation origin were classified for the first time in two types by D. Kuhlmann-Wilsdorf and N. Hansen [86] and subsequently studied experimentally by transmission electron microscopy in a series of studies by N. Hansen et al [87–89]. In particular, the cell boundaries were classified as random dislocation boundaries (JDB), formed by mutual intersection of the dislocations, and the boundaries of the blocks of the cells were defined as the geometrically necessary boundaries (GNB). This division was applied several years later by V.V. Rybin using the terms cell boundaries and knife-edge boundaries [90]. These boundaries divide the volumes of the elements in the dislocation structure, which are deformed by the combination of different sliding systems, and this is accompanied by a strain gradient. Shortly afterwards, the different types of the disoriented boundaries in the cellular substructure and the misorientation on them were investigated in detail in studies by the authors of this book (see, for example [52, 91]]. Later, J.G. Sevillano [92] introduced the concept of the geometrically necessary twins (GNT). At present, the entire spectrum of the geometrically necessary defects is used in the analysis of deformation mechanisms and relationships governing the formation of dislocation structures.

At usual grain sizes at the mesolevel $\rho_S > \rho_G$. For the FCC solid solutions this relationship is satisfied up to $d = 40$ μm [48]. With further refining of the average grain size at $d = 15$ μm ρ_G becomes greater than ρ_S [93]. At the same time, the critical grain size is in the vicinity of $<d> = 20$ μm. The same relationship $\rho_G > \rho_S$ can also be valid for the nanograins.

The component of the dislocation structure ρ_G was introduced by Ashby [3, 31] to evaluate the role of the grain boundaries in the formation of the dislocation structure. However, the inhibition of shear at the grain boundaries does not include all types of inhibition of the sliding dislocations at different boundaries. Different sub-boundaries, such as boundaries of the fragments, blocks and cells, inhibit the shear to a lesser extent than the grain boundaries but to a greater extent than the individual dislocations. This inhibition may provide a contribution to the component of the dislocation structure ρ_G. The first attempt to evaluate the role of the GNDs in the cellular dislocation substructure was made by H. Mughrabi [94, 95]. This problem is also investigated here.

Another type of inhibition of shear is the solid solution hardening. Although this type of hardening is less effective than the barrier inhibition, it is well-known that the effect of this inhibition increases the density of dislocations accumulated in the volume of the solid solution [96]. Nevertheless, no detailed analysis of the contributions of ρ_S and ρ_G to the total scalar density of the dislocations ρ in the solid solutions has been carried out. At the same time, according to the data obtained by the authors of this book [96], in deformation of the Cu–Mn solid solutions in the manganese concentration range from 0 to 25 at.%, the scalar density of dislocations, stored in the volume of the material, is determined mainly by the solid solution hardening τ_f. At present, it is not clear whether the solid solution hardening provides a contribution to the storage of the GND. This effect is studied in this section together with other defects, determined by the formation in the steel of cells, fragments, laths, etc.

The aim of this section is to discuss the role of different types of hardening in the formation of the density of geometrically necessary dislocations and compare the fraction of these dislocations with the statistically stored dislocations. The main types of hardening include: 1 – barrier inhibition of shear by different boundaries of grains, fragments, cells, martensite plates, laths, packets, etc in the steel, and 2 – solid solution hardening in the concentrated solid solutions.

Attention will be given to the available reports of the storage of dislocations in the plastic deformation of polycrystals of the FCC Cu–Mn solid solutions and a BCC steel with carbides and block hardening. The concentration of the Cu–Mn solid solutions, the composition of the steel, the density of microparticles in the steel, and the dimensions of the grains and other structural components of the investigated materials are presented in Table 3.3.

The deformation of the specimens of the Cu–Mn alloys and the steel was performed by tensile loading at room temperature. The strain rate in all cases was $\dot{\varepsilon} = 10^{-2}$ s^{-1}. The dislocation structure was studied by transmission diffraction electron microscopy using electron microscopes fitted with a goniometer, at an accelerating voltage of 125 kV. The scalar density of the dislocations ρ was determined by the well-known secant method, the density of the geometrically necessary dislocations was calculated using the equations (3.6) and (3.14). The value of χ in (3.14) was determined using the parameters

Table 3.3. Data for the investigated materials

Alloy	Alloy composition	Average grain size $<d>$ and size of other structural elements
Cu–Mn solid solutions	Cu+0.4 at.% Mn Cu+4.0 at.% Mn Cu+6.0 at.% Mn Cu+13.0 at.% Mn Cu+19.0 at.% Mn Cu+25.0 at.% Mn	$<d> = 60$ μm
Martensitic steel	0.34% C+0.85% Cr+2% Ni+0.5% Mo+0.5% V +balance Fe	Grain: $<d> = 34$ μm Plates: 0.9×2.4 μm Packets: 4×6 μm Laths: 0.2×6 μm Primary fragments: 92×640 nm Secondary fragments: 60×370 nm Cells: 30 nm

of the bending extinction contours on electron microscopic images (see, for example, [97]].

Attention will now be given to the mechanisms of inhibition of shear in the investigated materials. All the investigated materials are characterised by the presence of the polycrystalline contribution to the density of the GND (ρ_G). The role of the GND is determined in particular by the comparison of the stored scalar density of the dislocations ρ and the contribution of the density ρ_G to this density. In further stages, this comparison is carried out both directly and by analysis of the relationship ρ_G/ρ_S. In addition to polycrystalline hardening, the investigated material is also characterised by other hardening mechanisms. For example, the polycrystals of the Cu–Mn solid solutions contain the contribution of solid solution hardening which may result in the storage of both the component of the dislocation density ρ_S and ρ_G. At the same time, the Cu–Mn solid solutions are characterised by the formation of a cellular substructure (Fig. 3.31), which may contribute to both ρ_S and ρ_G. The BCC steels also contain the contributions of ρ_S and ρ_G. As a result of the martensitic transformation and subsequent annealing, the steels contain martensite plates, martensite packets of laths and, inside them, the dislocation fragments and cells (Figs. 3.32 and 3.33). The substructure contains the carbide particles (Fig. 3.33). In most cases, they are localised at crystal structure defects. The fragments in Fig. 3.33 with the network dislocation substructure are primary, with the cellular structure – secondary, and those with the dislocation

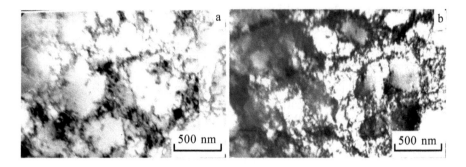

Fig. 3.31. Dislocation cellular substructure in the Cu+0.4 at.% Mn alloy at different strains: a – the non-misoriented cellular substructure at ε = 5%; b – the misoriented cellular substructure at ε = 30%.

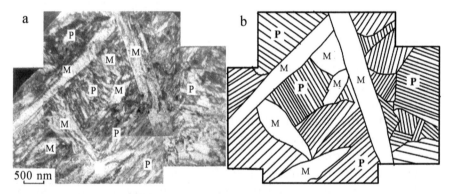

Fig. 3.32. Structure of the packet-plate martensite of the steel after quenching. P – the packets of lath martensite; M – martensite plates; a – micrograph, b – schematic representation.

free fragments – tertiary. In this sequence, they form in the steel either during annealing or deformation. Attention should be given to the fact that the three types the fragments in the steel with a different dislocation structure are similar to the three types of grains in the sub-microcrystalline materials, produced by severe plastic deformation [15, 46, 97]. It is surprising to note that the dislocation structure in the three types of grains is the same as in the three types of fragments. It should also be noted that in the deformation of the investigated materials, there is the entire set of the obstacles forming the contribution of ρ_G to the total scalar dislocation density ρ.

The total dislocation density ρ consists of the statistically stored dislocations with the density ρ_S and the geometrically necessary dislocations with the density ρ_G [3, 4, 31]. Therefore, $\rho = \rho_S + \rho_G$ in which ρ is the sum of these components. The experimental measured

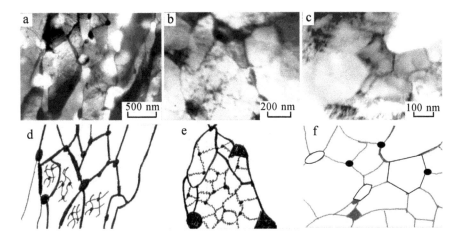

Fig. 3.33. Electron microscopic (a, b, c) and schematic (d, e, f) images of the primary (a, d), secondary (b, e) and test three (c, f) fragments, formed in deformation of the martensitic steel. The distribution of the carbide particles of the substructure is shown (● – particles of special carbides, ▬– cementite particles).

values of ρ, ρ_S and ρ_G for the Cu–Mn alloys are presented in Fig. 3.34. As indicated by Fig. 3.34, the magnitude of the contribution of ρ_G and ρ in the alloys of the Cu–Mn system is large. Although $\rho_S > \rho_G$, the contributions ρ_S and ρ_G are comparable at all concentrations of the solid solution. As indicated by Fig. 3.34, the component ρ_S rapidly increases in the strain range $\varepsilon = 0...0.05$. As the deformation proceeds, the behaviour of ρ_S and ρ_G is very similar, and ρ_S remains larger than ρ_G.

Figure 3.35 shows the dependences of ρ and the density of its components ρ_S and ρ_G, built-up in the deformation in different structural formations in the steel. These formations are – the grain, the martensite packet, the primary dislocation fragment, the martensite lath, and the secondary dislocation fragment. Their dimensions are presented in Table 3.3. Figures 3.34 and 3.35a can be compared. They have many common features: ρ_S and the beginning of plastic deformation ($\varepsilon < 5\%$) rapidly increases as in the case of ρ. The contribution of ρ_G in this case is smaller than the contribution of ρ_S. Further, with the development of deformation, ρ_G starts to 'catch up' with ρ_S. The difference between ρ_S and ρ_G is then maintained during further deformation. The behaviour of ρ_S and ρ_G, both in the polycrystals of the Cu–Mn alloys (Fig. 3.34), and in the polycrystals of the steel (Fig. 3.35a) has many similar features. In this case, the average grain size in the copper–manganese alloys does not depend

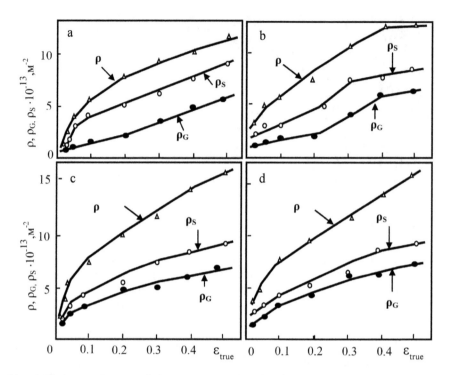

Fig. 3.34. Dependences of the average scalar density of dislocations ρ and its components $ρ_G$ and $ρ_S$ on the strain of the alloys of the Cu–Mn solid solutions at a different Mn concentration (at.%): a – 0.4; b – 6.0; c – 13.0; d – 19.0.

on the manganese concentration and is twice as high as in the steel (Table 3.3).

3.7. Storage of geometrically necessary dislocations and scalar dislocation density. The role of boundaries of different type

Attention will now be given to the role of structural components of a martensitic steel in the behaviour of the scalar (ρ) dislocation density and its components ($ρ_S$ and $ρ_G$) in relation to the strain [62] (Fig. 3.35). In analysis of the substructural elements of the steel special attention was given to their transverse dimensions.

Table 3.2 shows clearly that in the given range of the structural formations in the steel from the grains to the dislocation cells their size decreases by three orders of magnitude. In this case, Fig. 3.35, the average scalar density of dislocations slightly increases with strain. Large changes take place in the contributions of $ρ_S$ and $ρ_G$.

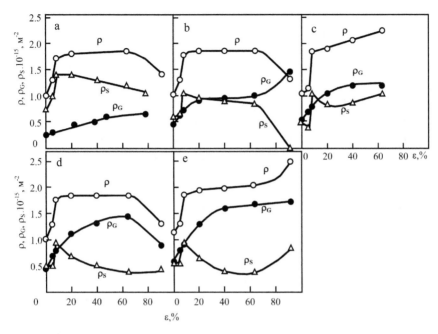

Fig. 3.35. Dependence of the average scalar density of dislocations ρ and its components ρ_G and ρ_S on the strain for different structural components of the steel: a – grain; b – martensite packet; c – primary fragment; d – martensite laths; e – secondary fragment.

Comparison of Fig. 3.35 and the data in Table 3.2 shows that with a decrease of the size of the substructural element, the component ρ_S decreases and the component ρ_G increases. Examination of the sequences in Figs. 3.35a–3.35e shows that with a decrease of the size of the structural element the scalar density of dislocations ρ slightly increases and the component ρ_G greatly increases which, in the final analysis, becomes dominant over ρ_S. The component ρ_S remains large to ε ≈ 10%, and with further deformation the contribution of ρ_S decreases with the refining of the structural element. As the density of the specific type of boundaries increases, their contribution to ρ_G increases and to ρ_S become smaller. We can examine the relative role of these contributions in the dependence on the path length of dislocations L with increase of the density of the boundaries, preventing shear. The path length of the dislocations L can be estimated using the data in Table 3.2. With a decrease of the size of the structural element the value of L decreases because it may exceed this size by no more than a factor of 3–5.

The main result is the fact that as the size of the structural formation in which the shear is inhibited decreases and also the dislocation path length L decreases, the contribution of ρ_G increases and that of ρ_S decreases. Although this result is not unexpected, because ρ_G is the density of the geometrically necessary dislocations, it has been obtained by experiments for the first time. This also determines the strong dependence of the GND on the size of the region in which the shear is inhibited. This effect takes place, regardless of the fact that a decrease of the size of the structural element changes the type of its boundaries, and the inhibition of dislocations at these boundaries becomes less effective.

The ratio ρ_G/ρ_S in dependence on the size of the structural element in the investigated steel is shown in Fig. 3.36. It shows that the ratio ρ_G/ρ_S in dependence of the size on the structural element l consists of two branches. The first branch at low l rapidly decreases downwards, the values of ρ_G/ρ_S in the second branch in the range of the values $l = 100$ nm–40 µm slowly decreases. The strong dependence of ρ_G/ρ_S on l forms in the range 30...150 nm and ends in the vicinity of 100...150 nm. Subsequently, the dependence ρ_G/ρ_S is almost horizontal. The dotted line in Fig. 3.36a shows the critical size, ~100

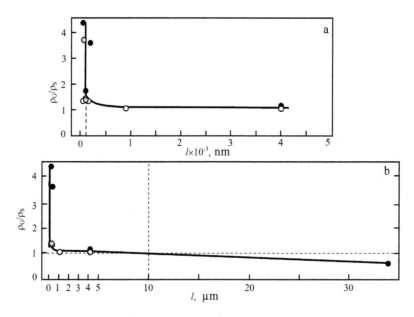

Fig. 3.36. The ratio ρ_G/ρ_S for different structural elements of the steel: a – for the size range 0...1000 nm, b – for the range 0...40 µm. The scale of the figures a and b on the size axis differs by a factor of 5.

nm. This corresponds to the critical size of the grain of 100 nm in the ultrafine-grained polycrystals [15]. At the grain size of $d < 100$ nm it becomes dislocation-free. The density of dislocations ρ_G becomes a conventional value. It reflects the presence of the dislocations and disclinations of the grain boundaries and disclinations in the ternary junctions [70].

The critical size $d = 100$ nm is also found in work hardening [98, 99]. In the vicinity of $d = 100$ nm, the strain hardening coefficients in the linear stages II and IV, θ_{II} and θ_{IV} rapidly decrease and with a further increase of the grain size their values correspond to the values of the polycrystals at the mesolevel. The point is that as indicated by the experimental data, in the vicinity of $d = 100$ nm, the increase of the grain size is accompanied by a change of the dominant mechanisms of deformation of the nanocrystals [36, 98, 99]. These mechanisms with increasing grain size transform from grain boundary–diffusion to the grain boundary–dislocation and lattice–diffusion and then to lattice–dislocation. Figure 3.36b shows the behaviour of ρ_G/ρ_S at large dimensions of the structural element. At $l \sim 100$ μm $\rho_G = \rho_S$ and the ratio $\rho_G/\rho_S = 1$. This is the second critical size above which the component ρ_S becomes larger than ρ_G. The ratio $\rho_G/\rho_S < 1$ is typical of the components of the dislocation structure in the mesoregion.

3.8. Concentration dependence of the main parameters of the dislocation structure in the FCC solid solutions

This effect can be investigated using the data obtained for the Cu–Mn alloys [100]. The concentration dependences of ρ, ρ_S and ρ_G are shown in Fig. 3.37a. Measurements were taken for $\varepsilon = 5\%$. The scalar dislocation density ρ rapidly increases with increasing manganese concentration. This is associated with the increase of the degree of solid solution hardening τ_f. The dependence of τ_f on the manganese concentration is shown in Figure 3.37b. The data presented in Fig. 3.37 indicate the obvious role of solid solution hardening in the buildup of dislocations. The component ρ_G increases. In the investigated manganese concentration range the value of this component is almost doubled. On the other hand, the contribution of ρ_S in this concentration range increases only slowly and only to the composition of 10 at.% Mn (Fig. 3.37a). In subsequent stages, the value of ρ_S remains constant with increasing manganese concentration. Undoubtedly, the increase of the scalar dislocation

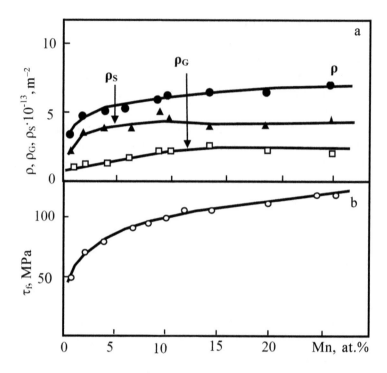

Fig. 3.37. Concentration dependences of the scalar (ρ) dislocation density and its components (ρ_S and ρ_G) (a) and the solid solution hardening τ_f (b). The data are given for the deformation temperature $T = 293$ K, strain $\varepsilon = 5\%$.

density ρ in the investigated range of the manganese concentration is caused by the increase of ρ_G. The constant value of ρ_S in the same manganese concentration range is associated with the retention of the properties of the individual dislocations, regardless of the change of the composition of the solid solution. It is well-known that the stacking fault energy in the investigated manganese concentration range changes only slightly [101]. It may therefore be expected that the strength of the dislocation barriers does not change with increasing manganese concentration. The increase of ρ_G is caused by the large increase of the deformation gradient with increasing solid solution hardening.

3.9. Cellular substructure: dislocation density ρ_S and ρ_G and the cell size

The dislocation substructure in the manganese concentration range

0...8 at.% is cellular (Fig. 3.31). With a further increase of the solid solution concentration the substructure changes to the cellular-network and remains in this state to the manganese concentration of 25 at.%. In a martensitic tempered steel, discussed above, the substructure is more complicated. This substructure retains the martensite plates, the packets of laths and a complicated dislocation structure – fragmented and non-fragmented – and also contains the dislocation cells and networks. Examples of the substructure of the steel are shown in Fig. 3.33. A special role in both the Cu–Mn solid solutions and in the steel is played by the cellular substructure.

The main components in the cellular substructure forming the dislocation density ρ are the dislocation density in the walls of the cells ρ_w and the dislocation density inside the cells ρ_{in}. They are linked together by the following relationship:

$$\rho = \rho_w \cdot f_w + \rho_{in} f_{in} \qquad (3.27)$$

where f_w and f_{in} are the volume fractions of the material of the walls and the body of the cell, respectively. Experimental data show that in the cells situated in the substructure of a copper–manganese alloy, $\rho_S > \rho_G$. On the other hand, in the cells situated in the fragmented substructure of the investigated steel $\rho_G > \rho_S$. When comparing the size of the cells in the Cu + 8 at.% Mn alloy in which $D =$ 300–500 nm and in the steel where $D = 30$ nm. It is evident that the size of the cells in these materials differs by an order of magnitude. It is well-known that the walls of the cells inhibit dislocation shear. Therefore, it is not surprising that for the steel $\rho_G > \rho_S$, and in the Cu+8 at.% Mn alloy $\rho_S > \rho_G$.

In the previous section attention was given to the storage of the scalar dislocation density and its components in the plastic deformation of the FCC Cu–Mn solid solutions and a tempered BCC martensitic steel. Several results will now be mentioned. It is well-known that the dislocations carrying out deformation in forming substructures in the deformed material can be divided into two types on the basis of the physics of their origin. The first type includes the dislocations producing uniform deformation under the effect of applied stress. They are referred to as the statistically stored dislocations. Their density is denoted by ρ_S. The second type is caused by the inhomogeneity of deformation caused by local obstacles and local stress concentrators. These dislocations are referred to as geometrically necessary dislocations. Their density is ρ_G.

In this section, attention is given to the data obtained in direct electron microscopic measurements carried out to determine the total dislocation density ρ. Using the results of measurements of the curvature–torsion of the crystal lattice χ the density of the geometrically necessary dislocations ρ_G was also determined. The density of the statistically stored dislocations was determined as $\rho_S = \rho - \rho_G$. The value of ρ_G, determined by the inhibition of the dislocations at different boundaries was measured: 1. grain boundaries, 2. the boundaries of martensite packets, 3. the boundaries of martensite laths, 4. the boundaries of dislocation fragments, 5. the cell boundaries. The results show that the ρ_G/ρ_S ratio depends on the value of the closed structural element whose boundaries inhibit shear. The value ρ_G/ρ_S increases with a decrease of the shear strain. The critical size of the substructural element, equal to 100 nm, was determined. When the size is reached, the dependence ρ_G/ρ_S rapidly changes in its form. It is important to note the fact that the value 100 nm also corresponds to the size of the dislocation-free cells in the nanomaterials (the second critical grain size).

Attention has also been paid to the storage of dislocations under the effect of solid solution hardening. The experimental data show the important role of the component ρ_G in the storage of dislocations in the solid solutions. The effect of the size of the dislocation cells on the ratio of the components of the dislocation density ρ_G and ρ_S was investigated. The results show that the decrease of the cell size increases ρ_G in comparison with ρ_S.

Experiments carried out in [59, 62] clearly demonstrated the important role played in the quantitative relationships by the geometrically necessary dislocations in the formation of dislocation substructures in the conditions of the effect of different hardening mechanisms.

References

1.   Friedel G., Dislocations, Moscow, Mir, 1967..
2.   Seeger A., A sliding mechanism and strengthening in the face-centered cubic and hexagonal close-packed metals, in: Dislocations and mechanical properties of crystals. Moscow, IL, 1960, 179–289.
3.   Ashby M.F., *Phil. Mag.*, 1970, V. 21, 399–424.
4.   Courtney T.H., Mechanical behavior of materials, International Editions, McGraw - Hill, 2000.

5.   Hirt J., Lote I., Dislocation theory, Moscow, Atomizdat, 1972.
6.   Koneva N.A. et al., in : New methods in physics and solid mechanics. Part I. (Ed. V.E. Panin). Tomsk, TSU Publishing House, 1990, 83–93.
7.   Koneva N.A., Internal long-range stress fields in ultrafine grained materials, Severe plastic deformation. Toward bulk production of nanostructured materials. (Ed. B.S. Altan), New York, Nova Science Publishers, Inc., 2005, 249–274.
8.   Kubin L.P., Mortcusen A., *Scr. Mat.,* 2003, V. 48, 119–125.
9.   Gao H., Huang Y., *Scr. Mat.,* 2003, V. 48, 113–118.
10.   Koneva, N.A., et al., in: Physics of defects of the surface layers of materials (ed. A.E. Romanov), Leningrad, A.F. Ioffe Institute, 1989, 113–131.
11.   Koneva N.A., et al., *Mat. Sci. Eng.*, 2001, V. A319–321, 156–159.
12.   El-Dasher B.S., et al., *Scr. Mat., 2003,* V. 48, 141–145.
13.   Conrad H., Strain hardening model to explain the influence of grain size on the flow stress of metals, in: Ultrafine grains in metals, Coll. articles, (Translated from English, Ed. L. Gordienko), Moscow, Metallurgiya, 1973, 206–219.
14.   Koneva, N.A., Kozlov E.V., in: Structural levels of plastic deformation and fracture. (Ed. V.E. Panin), Novosibirsk, Nauka, Sib. branch, 1990, 123–186.
15.   Kozlov E.V., et al., *Fiz. mezomekhanika*, 2004, V. 7, No. 4, 93–113.
16.   Sharkeev Yu.P., et al., *Fiz. Met. Metalloved.*, 1985, V. 60. No. 4, 816–821.
17.   Valiev R.Z., et al., *Acta Met.,* 1994, V. 42, No. 7, 2167–2475.
18.   Kozlov E.V., Structure and resistance to deformation of UFG metals and alloys, in: Severe plastic deformation. Toward bulk production of nanostructured materials / Ed. B.S. Altan, New York, Nova Science Publishers, Inc., 2005, 295–332.
19.   Kozlov E.F., et al., *Fundamental problems of modern materials*, 2009, V. 6, No. 1, 7–11.
20.   Kurdyumov G.V., et al., Transformations in the iron and steel, Moscow, Nauka, 1977.
21.   Bhadesia H.K., Bainite in steels, London, Institute of Materials, 1992.
22.   Valiev R.Z., Langdon T.G., *Progr. Mat. Sci.,* 2006, V. 51, 881–981.
23.   Mulyukov R.R., Noskov N., Submicrocrystalline metals and alloys, Ekaterinburg, Ural Branch of Russian Academy of Sciences, 2003.
24.   Kozlov E.V., et al., *Fiz. mezomekhanika,* 2009, V. 12, No. 4, 93–106.
25.   Kozlov E.V., et al., *Fundamental problems of modern metallurgy,* 2009, V. 6, No. 2, 14–24.
26.   Koneva N.A., et al., *ibid,* 2010, V. 7, No. 1, 64–70.
27.   Koneva N.A., et al., *Mat. Sci. Forum*, 2010, V. 633–634, 121–128.
28.   Koneva N.A., et al., *Bulletin of the Russian Academy of Sciences, Physics,* 2010, V. 74, No. 5, 592–596.
29.   Kuhlmann-Wilsdorf D., *Phil. Mag.,* 1999, V. 79, No. 4, 955–1008.
30.   Kubin L.P., et al., in: Dislocations in Solids. V. 11. (Ed. F.N.R. Nabarro and M.S. Duesbery), Amsterdam–Tokyo, Elsevier, 2002, 101–192.
31.   Ashby M.F., in: Strengthening Methods in Crystals, London, Science Publishers Ltd, 1971, 137–190.
32.   Koneva, N.A., et al., *Fundamental problems of modern materials*, 2010, V. 8, No. 1, 80–86.
33.   Orlov A.N., *Fiz. Met. Metalloved.,* 1977, V. 44, No. 5, 966–970.
34.   Kozlov E.V., et al., *Pis'ma o materialakh,* 2011, V. 14, No. 3, 15–18.
35.   Kozlov E.V., et al., *Fiz. mezomekhanika,* 2007, V. 10, No. 3, 81–92.
36.   Kozlov E.V., et al., *Fiz. mezomekhanika,* 2006, V. 9, No. 3, 81–92.
37.   Kozlov E.V., et al., *Ann. Chim. Fr.*, 1996, V. 21, 427–442.

38. Kozlov E.V., et al. Structural evolution of ultrafine grained copper and copper-based alloy during plastic deformation, in: Ultrafine grained materials, II (eds. Y.T. Zhu, T.G. Langdon, R.Z. Mishra et al.) USA, TMS publication, 2002, 419–428.

39. Hall E.O., *Proc. Phys. Soc.*, 1951, V. 64B, 747–753.

40. Petch N.J., *J. Iron Steel Inst.*, 1953, V. 174, 25–28.

41. Kozlov E.V., et al., *Mat. Sci. Forum*, 2008, V. 584–586, 35–40.

42. Segal V.M., et al., Processes of plastic structure formation in metals, Minsk, Nauka i tekhnika, 1994.

43. Valiev R.Z., Alexandrov I.V., Nanostructured materials produced by severe plastic deformation, Moscow, Logos, 2000.

44. Pantleon W., *Scr. Mat.*, 2008, V. 58, 994–997.

45. Koneva N.A., et al., in: New methods in physics and solid mechanics. Part 1 / (Ed. V.E. Panin), Tomsk, TSU Publishing House, 1990, 83–93.

46. Koneva N.A., et al., *Izv. AN, Ser. Fiz.*, 1998, V. 62, No. 7, 1350–1356.

47. Koneva N.A., et al., *Izv. VUZ Fizika*, 2014, V. 57, No. 2, 45–53.

48. Koneva N.A., et al., in: Physics of defects of the surface layers of materials (ed. A.E. Romanov), Leningrad, A.F. Ioffe Institute, 1989, 113–131.

49. Sharkeev Yu.P., et al., *Fiz. Met. Metalloved.*, 1985, V. 60, No. 4, 815–821.

50. Perevalova O.B., Koneva, N.A., *ibid*, 2003, V. 98, No. 4, 106–112.

51. Popov N.A., Lapsker I.A., On the role of deformation and annealing twins in the formation of mechanical properties of FCC alloys, Plastic deformation of alloys (eds. L.E. Popov and N.A. Koneva), Tomsk, TSU, 1986, 241–248.

52. Kozlov E.V., et al. *Mat. Sci. Eng.*, 2001, V. A319–321, 261–265.

53. Cahn R.W., The coming of materials science, Amsterdam–Tokyo, Elsevier Science, 2001.

54. Kozlov E.V., et al., *Fiz. mezomekhanika*, 2011, V. 14, No. 3, 95–110.

55. Li J.C.M., *Metall. Soc. of AIME*, 1963, V. 227, 239–347.

56. Orlov L.G., *Fiz. Tverd. Tela*, 1967, V. 9, No. 8, 2345–2349.

57. Valiev R.Z., et al., *Metallofizika*, 1983, V. 5, No. 2, 94–100.

58. Koneva N.A., Kozlov E.V., *Izv. VUZ, Fizika*, 1990, No. 2, 89–106.

59. Koneva N.A., et al., *ibid*, 2014, V. 57, No. 2. 45–53.

60. Koneva N.A., et al., *Izv. RAN, Ser. Fiz.*, 2010, V. 74, No. 5, 630–634.

61. Koneva N.A., Kozlov E.V., in: Advanced Materials III, (ed. D.L. Meerson), Moscow Institute of Steel and Alloys, 2009, 55–140.

62. Koneva N.A., et al., *Izv. VUZ, Fizika*, 2009, V. 52, No. 9/2, 5–14.

63. Hirsch P., et al., Electron microscopy of thin crystals, Moscow, Mir, 1968.

64. Saltykov S.A., Stereometric metallography, Moscow, Metallurgiya, 1970.

65. Kozlov E.V., et al., *Izv. RAN, Ser. Fiz.*, 2009, V. 73, No. 9, 1295–1301.

66. Hansen N., *Acta Met.*, 1977, V. 25, 863–869.

67. Hansen N., *Metall. Trans.*, A, 1985, V. 16A, 2167–2190.

68. Andrievsky R.A., *Rossiiskie Nanotekhnologii*, 2011, V. 6., No. 5, 34–42.

69. Koneva N.A., et al., *Mat. Sci. Forum*, 2010, V. 633–634, 605–611.

70. Koneva N.A., et al., *Mat. Sci. Forum*, 2008, V. 584–586, 269–274.

71. Kozlov E.V., et al., Deformatsiya i razrushenie materialov, 2009, No. 6, 22–27.

72. Koneva N.A., et al., in: Structure, phase transformations and properties of nanocrystalline alloys, (ed. Noskova N.I.), Ekaterinburg, IMP UB RAS, 1997, 125–140.

73. Morris D.G., Morris M.A., *Acta Met.*, 1991, V. 39, No. 8, 1763–1770.

74. Martin J.W., et al., Stability of microstructure in metallic systems, Cambridge University Press, 1997.

75. Koneva N.A., et al., in: Problems of nanocrystalline materials (ed. V. Ustinov and

N.I. Noskova), Ekaterinburg, Ural Branch of Russian Academy of Sciences, 2002, 57–71.

76.   Jones A.R., Hansen N., *Acta Met.*, 1981, V. 29, No. 3, 509–599.

77.   Hunderi O., Ryum N., *Acta Met.*, 1981, V. 29, No. 10, 1737–1745.

78.   Koul A.K., Pickering F.B., *Acta Met.*, 1982, V. 30, No. 9, 1303–1308.

79.   Tweed C.J., et al., *Acta Met.,* 1984, V. 32, No. 9, 1407–1414.

80.   Wörner S.N., Cabo A., *Scripta Met.*, 1984, V. 18, 565–568.

81.   Van Vleck L.X., Microstructure, Cahn R.W., Physical metallurgy, Vol II. Phase transformations. Metallography. Moscow, Mir, 1968, 402–430.

82.   Anderson M.P., et al., *Acta Met.,* 1984, V. 32, No. 8, 783–791.

83.   Srolovitz D.J., et al., *Acta Met.*, 1984, V. 32, No. 8, 793–802.

84.   Srolovitz D.J., et al., *Acta Met.*, 1984, V. 32, No. 9, 1429–1438.

85.   Kozlov E.V., et al., *Izv. VUZ, Chernaya metallurgiya*, 1999, No. 8, 35–39.

86.   Kuhlmann-Wilsdorf D., Hansen N., *Scr. Met. Mater.*, 1991, V. 25, 1557–1562.

87.   Liu Q., Hansen N., *Scr. Metall. Mater.*, 1995, V. 32, 1289–1295.

88.   Liu Q., et al., *Acta Mater.*, 1998, V. 46, No. 16, 5819–5838.

89.   Hughes D.A., Hansen N., *Acta Mater.*, 2000, V. 48, 2985–3004.

90.   Rybin V.V., Large plastic deformation and fracture of metals, Moscow, Metallurgiya, 1986.

91.   Koneva, N.A., Kozlov E.V., *Izv. VUZ, Fizika*, 1991, No. 3, 56–70.

92.   Sevillano J.G., *Scr. Mat.,* 2008, V. 59, 135–138.

93.   Dudarev V.E., *Izv. VUZ, Fizika,* 1991, No. 3, 35–46.

94.   Mughrabi H., *Mat. Sci. Eng*, 2001, V. A319–321, 139–143.

95.   Mughrabi H., *Acta Mat.,* 2006, V. 54, 3417–3427.

96.   Trishkina L.I., Koneva N.A., Influence of doping on the evolution of the substructure with deformation of solid solutions based on Cu–Al and Cu–Mn, in: Natural Sciences and Humanities in the XXI century (ed. N.A. Koneva, et al.). Tomsk: ASU Publishers, 2004, 25–33.

97.   Koneva, N.A., Kozlov E.V., in: Structural and phase states and properties of metallic systems (ed. Potekaeva A.I.), Tomsk, Publishing house of the NTL, 2004, 83–110.

98.   Koneva N.A., et al., in: Advanced Materials and Technologies. Proceedings of the Regional scientific-technical conference. Tomsk, Publishing house Print Manufactory, 2009, 8–18.

99.   Koneva N.A., Kozlov E.V., *Materials Science Forum*, 2011, V. 683, 183–187.

100. Kozlov E.V., et al., *Izv. RAN, Ser. Fizicheskaya*, 2011, V. 75, No. 5, 713–715.

101. Steffens Th., et al., *Phil. Mag. A*, 1987, V. 56, No. 2, 161–173.

4

Dislocation structure and internal stress fields

4.1. Introduction

The long-range internal stress fields (IS) play an important role in different processes taking place in ultrafine-grained materials produced by severe plastic deformation (SPD). Insufficient attention has been paid to these fields in experiments. Transmission electron microscopy (TEM) is the most suitable method for studying the internal stress fields. In this chapter, attention is given to the methods of electron microscopic examination of the internal stress fields. The results of examination of the structure of the grains and subgrains and of their boundaries in ultrafine-grained copper and nickel are presented. The sources of the internal stress fields are outlined. The experimental data for the distribution of the internal stress fields are used to describe the structure of the grains of the ultrafine-grained material.

The problem of the elastic long-range internal stress fields, formed in the deformed metals and alloys has been in the centre of attention of researchers for many years (see, for example [1–8]). This is caused by a number of reasons. The internal stresses provide a significant contribution to the deformation resistance for the formation and evolution of the defective structure and its rearrangement during deformation is closely linked with the relaxation of internal stresses. The nucleation and propagation of microcracks is also determined by the relaxation of the internal stresses. Finally, internal stresses play a significant role in phase transitions.

In particular, the internal stresses are defined on the basis of the region in which they form. Consequently, there are macro-, meso- and microstresses. The macrostresses are localised in the entire specimen or in a large part of its volume, the mesoscopic internal stresses are localised in the volumes with the length of several to hundreds of microns. In the polycrystals, the mesostresses are localised in the volume of several grains or in part of the grain volume. The fields of the internal microstresses are localised in sections with the size of several microns or less.

The meso- and microfields of the stresses have been studied insufficiently by experiments. This applies in particular to the ultrafine-grained materials. At the same time, in the case of these materials it is important to understand the role of internal stresses in different processes which take place in these materials. Examination of the nature of the fields of internal stresses in the ultrafine-grained materials is an important task because the elastic energy stored in them determines the instability of the structure of these materials, especially with increasing temperature [9]. Over a number of years, the authors of this book and their colleagues have carried out experimental studies of the internal stresses in the ultrafine-grained metals and alloys by transmission electron microscopy. In this chapter, these investigations are generalised together with the analysis of the literature data for this problem.

4.2. Methods for measuring internal stresses

The meso- and microstresses, formed in deformation in the metallic materials, are often analysed by X-ray diffraction analysis. When using X-ray diffraction analysis, the amplitude of the internal stresses is evaluated using the broadening of the X-ray interferences [10–13]. Here the application of the transmission electron microscopy makes it possible to examine the internal stresses in greater detail, measure both the meso- and microfields, investigate their distribution in the volume of the material and in individual grains of the polycrystal, and identify the sources of the internal stresses. Taking these considerations into account, it may be assumed that transmission electron microscopy is the principal method of studying the internal stresses in the ultrafine-grained materials.

When using transmission electron microscopy, information on the internal stresses can be obtained: 1) by measuring the radius of bending of the free dislocation segments [1, 14–16] and 2) measuring the parameters of the bending extinction contours [5, 14, 17–20]. The two methods will be examined in greater detail.

The first method can be used to determine the stress distribution in the volume of the material. The components (τ_e) of the elastic fields which can be determined using the data on the curvature of the dislocation lines are the cleavage stresses. The radius (r) of the curvature of the dislocation line is linked with τ_e by the relationship:

$$\tau_e = \frac{Gb}{2r},$$
(4.1)

where G is the shear modulus, b is the Burgers vector. Radius (r) of curvature of the dislocations can be determined if the height of the segments of its arc and the distance between the dislocation pinning points are known. Figure 4.1 shows the electron microscopic image of a grain of ultrafine-grained copper. The arrows indicate the dislocations which can be used to determine r and subsequently τ_e. The results of measurements of r must be averaged out taking into account the scatter of the angles between the plane of the foil and the plane in which the dislocations are situated.

The second method is based on determining the bending–torsion of the crystal lattice using the bending extinction contours observed on the electron microscopic images of the structure of the material. Bending–torsion forms when the thickness of the foil is decreased as a result of the presence of internal stresses in the thick material

Fig. 4.1. The method for determining the amplitude of cleavage stresses τ_e from the curvature of the dislocation lines. The bent dislocations are indicated by the arrows.

from which the foil is produced. The material 'remembers' the stress fields and undergoes bending–torsion during thinning. The pattern of the bending extinction contours reflects the bending–torsion of the crystal lattice in the material [21]. The bending–torsion of the crystal lattice can be estimated quantitatively either by the mean radius (R) of curvature–torsion or by the components of the tensor of the curvature–torsion χ_{ij}:

$$1/R = \chi_{ij} = \partial\varphi / \partial l, \tag{4.2}$$

where $\partial\varphi / \partial l$ is the gradient of continuous misorientation. The characteristics included in (4.2) in the crystal are local. The bending extinction contour is localised in the section with the same orientation of the reflecting plane. This intensity decreases from the centre to the edges with increase of the distance from the positions of the accurate Bragg reflection. The continuous misorientation gradient is normal to the contour line. The types of deformation of the crystal lattice – bending, torsion or the mixed case – are identified on the basis of the mutual orientation of the lines of the extinction contour and the vector of the acting reflection $\vec{g}$. The diagram of this phenomenon and the method of identifying are presented in Fig. 4.2. The bending of the crystal lattice in this figure is generated by the edge dislocations (Fig. 4.2a), torsion by screw dislocations (Fig. 4.2b). The atomic planes in the crystal are normal to the vector of acting reflection $\vec{g}$. If the normal $\vec{n}$ to the contour line is parallel to the vector of the acting reflection ($\vec{n}$, $\vec{g}$), the reflecting planes are inclined or the bending of the crystal lattice takes place (Fig. 4.2

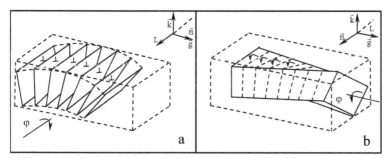

Fig. 4.2. The diagram of bending (a) and torsion (b) of the crystal lattice in the local areas of the foil: L – the line of the bending extinction contour; $\vec{g}$ – the direction of acting stresses, $\vec{n}$ – the direction of the normal to the line of the extinction contour, $\vec{k}$ – the direction of the incident electron beam, φ – the angle of inclinations of the goniometer. The dislocations causing bending (b) and torsion (b) are indicated.

a). In this case, the non-diagonal components of the bending–torsion tensor are determined. If $\vec{n}$ is perpendicular to $\vec{g}$ ($\vec{n} \perp \vec{g}$) the torsion of the reflecting planes of the lattice takes place (Fig. 4.2b). In this case the diagonal components of the tensor χ are determined. If $\vec{n}$ and $\vec{g}$ form some intermediate angle, the mixed case of bending–torsion is observed.

In determining the bending–torsion of the crystal lattice measurements are taken either of the speed of displacement of the bending extinction contour with the variation of the angle (φ) of the inclination of the goniometer or of the width of the contour. Special experiments with simultaneous application of both methods show that the width of the contour in terms of misorientation for the metals of the first long period of the Periodic table of elements equals approximately 1° at an accelerating voltage of 100–150 kV

If bending–torsion takes place in the investigated section of the crystal and there are no dislocations, the elastic bending–torsion takes place. Dislocation bending–torsion (Fig. 4.2) is produced by the local excess dislocation density: $\rho_\pm = \rho_+ - \rho_-$. Here ρ_+ and ρ_- are the densities of the dislocations of different sign. The following relationship holds in the latter case

$$\rho_\pm = \frac{1}{b}\frac{\partial \varphi}{\partial l} = \frac{\chi}{b}. \tag{4.3}$$

In dislocation bending–torsion the scalar dislocation density $\rho = \rho_+ + \rho_-$ should not be lower than the excess density determined from equation (4.3). If the scalar dislocation density ρ measured locally is lower than the values of $\rho_\pm$ calculated from equation (4.3), i.e. $\rho < \rho_\pm$, in this case the elastic bending–torsion is superimposed on dislocation bending. This bending–torsion will be called elastic–dislocation. In this case, the value $\rho_\pm$ is conventional because according to the definition it cannot ever exceed the values of ρ. The method described above can be used to restore several components of the tensor of the of the stresses of the internal field in the crystal, determine the sources of the internal stresses, measure their amplitude (χ) and determine the laws of decrease of the amplitude with increase of the distance from the source of internal stresses [17, 19].

Figure 4.3 shows an example of the bending extinction contour which can be used to measure χ. The source of the internal field in Fig. 4.3 is the particle (P) situated at the grain boundary (Gb) in ultrafine-grained copper. Grain A contains only the elastic bending–

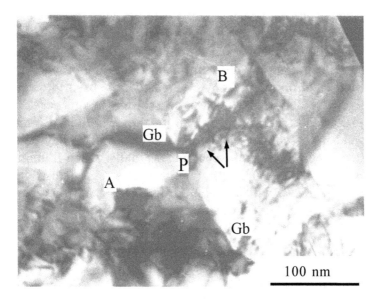

100 nm

Fig. 4.3. The bending contours from the particle (P) of Cu$_3$Sn at the boundary of the grains A and B. Gb is the grain boundary.

torsion (dislocations in this grain are not found), whereas grain B is characterised by elastic-dislocation bending–torsion. Measurements of the width of the extinction contour give the following average values of χ: in the grain A $\chi = (2.1\pm1.1) \cdot 10^6$ m^{-1}, in grain B $\chi = (0.9\pm0.2) \cdot 10^6$ m^{-1}. It may be seen that with increase of the distance from the grain boundary the contour widens (as shown by the arrows in Fig. 4.3b) indicating a decrease of the field amplitude.

Using the values of χ it is possible to determine the moment stresses τ_e. Transition from χ to τ_e takes place using the following relationships [14, 22, 23]: (1) in the case of elastic bending–torsion:

$$\tau_e = Gt \qquad (4.4)$$

(t is the thickness of the investigated specimen in the electron microscope) and (2) in the case of dislocation bending–torsion

$$\tau_e = Gt\chi / \pi(1-\nu) \qquad (4.5)$$

or

$$\tau_e = \alpha_e G\sqrt{b\chi} \qquad (4.6)$$

(ν is the Poisson coefficient, α_c is a coefficient which depends on the type of dislocation ensemble [15]). The relationship (4.4) corresponds to the fields of stresses of a relaxed pile up or a shear zone. The relationship (4.6) corresponds to the stress fields inside a dislocation charge. The value α_c changes in the range 0.1–1.5 [15, 23].

When determining τ_e it is important to remember that the transmission electron microscopy methods can be used to determine only the residual internal stresses. In addition to this, as a result of the relaxation structure in unloading the specimen prior to its thinning and in the course of thinning the values of τ_c may be lower than those found in the thick specimens. Finally, when thinning the specimen part of the sources of the internal stress fields, situated close to the surface of the specimen, disappears and the spectrum of the internal field can change to a certain extent.

4.3. Structure of ultrafine-grained metals and alloys

Structure of the micrograins

The grains of the metals and alloys can be efficiently refined by severe plastic deformation (SPD) [24–26]. There are a number of SPD methods. It may be multiple compression and pressing, rolling, hydraulic extrusion, etc. The structure of the grains and subgrains produced as a result of the SPD has the form of a suddenly arrested evolution of the ensemble of the dislocations, sub-boundaries and high-angle boundaries. The rate of the microprocesses of deformation and relaxation in different areas of the material may vary. Therefore, the ultrafine-grained material represents a spectrum of different structural states. The internal structure of the grains has not been studied sufficiently by experiments. At the same time, as mentioned previously, the grains of the ultrafine-grained materials are characterised by a complicated internal structure [27], the types of boundaries greatly differ, and there is the anisotropy of the grains and subgrains [28]. The anisotropy of the grains may be reduced, for example, in multiple passage in the conditions of equal channel angular pressing (ECAP) [29].

The structure of the materials with the submicron grain size has been studied extensively. One of the most important problems in these investigations is the determination of the true grain size. This is carried out using the methods of transmission electron microscopy and X-ray diffraction analysis. The determination of the true grain size is a relatively complicated task. The results, obtained by X-ray diffraction analysis, are strongly affected by the fields of internal stresses, present in the ultrafine-grained materials. The interpretation

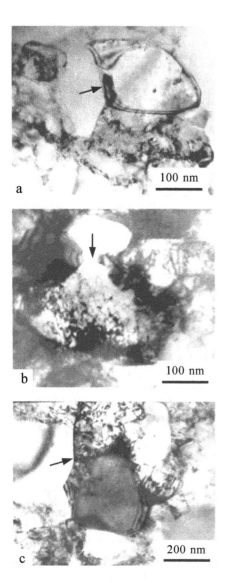

Fig. 4.4. Examples of three types of grains in ultrafine-grained copper: the grains free from dislocations (a), the grains containing chaotically distributed dislocations, forming networks (b), the grains containing dislocation cells and/ or fragments (c). The appropriate types of grains are indicated by the arrows.

of the transmission electron microscopy data is difficult because of the complicated contrast of different boundaries and sub-boundaries and as a result of the different structure of the grains of the ultrafine metals which may contain the cells and fragments (sub-grains) having different dislocation boundaries.

The investigation of the authors of this book with colleagues [14, 27, 28, 30] carried out on ultrafine-copper and nickel, prepared by ECAP and torsion under quasi-hydrostatic pressure (THP) followed by annealing for stabilisation of the structure, showed that the grain structure of the ultrafine-grained metals is characterised by two special features. The first special feature is as follows. The experimental results show that the grains of different sizes have different dislocation structures. There are three types of grains: 1. The grains not containing a dislocation; 2. The grains containing dislocations (the dislocations in these grains are distributed chaotically, or form dislocation networks), and 3. The grains containing dislocation cells or fragments (subgrains). The average sizes of these three subgrain types increase in the sequence described above. Figure 4.4 shows the examples of electron microscopic images of the three subgrains for the ultrafine-grained copper.

The second special feature of ultrafine metals is the presence of a wide distribution of the grain size of the grains. Figure 4.5 shows the corresponding distribution for the ultrafine-grained copper and nickel. In these distributions, the different crosshatching indicates the fraction of the grains with different defective structures. Analysis of the data, presented in Fig. 4.5, shows the mechanism of micrograins structure in SPD. The deformation is accompanied by the storage of a high density of dislocation inside the grains and, subsequently, a cellular structure forms inside the grains and then the fragmented structure. The dislocation density in the boundaries of the fragments and the misorientation angle at the boundaries increase. The fragment boundaries transform to new high-angle grain boundaries. Subsequently, the entire cycle is repeated during deformation. This indicates that inside the micrograins there is the usual cycle of evolution of the dislocation structure in the low-energy branch of the transformations in it [31]: the chaotic distribution of the dislocations → dislocation tangles → cells → fragments → new grains. Intermediate annealing accelerates the process of evolution of the dislocation structure. Previously, the important role of the processes of fragmentation in the course of formation of the grains of the submicron size was stressed in [32].

The types of boundaries

A number of models of the structure of the specific grains of the ultrafine-grained metals have been proposed [26, 32–35]. Special attention is paid to the non-equilibrium nature of the grain boundaries. The grain boundaries in the ultrafine-grained metals are distorted, containing lattice dislocations of different systems and are the sources of internal stresses. Consequently, they are characterised

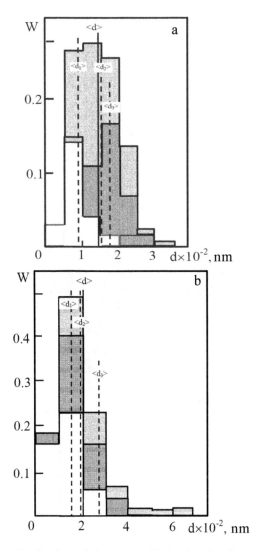

Fig. 4.5. The size distribution of the grains (*d*) in the ultrafine-grained nickel (a) and ultrafine-grained copper (b). Ultrafine-grained nickel was prepared by THP, ultrafine-grained copper by ECAP. The results of measurements for the ultrafine-grained nickel are presented for the sections situated at a distance of 5 mm from the axis of rotation. The bright areas of distribution — the fraction of the dislocation-free grains, the crosshatching ⬚ indicates the fraction of the grains with the dislocations (chaos and dislocation networks), ⬚ — the fraction of the grains with the cells and fragments; ⬚ — the fraction of the recrystallised grains; FINISH CAPTION

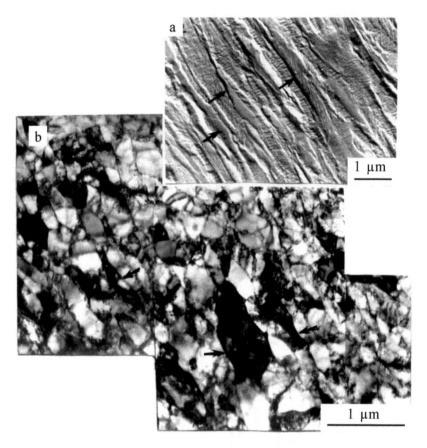

Fig. 4.6. Electron microscopic images of the grain structure of the ultrafine-grained nickel prepared by equal channel angular pressing on the replicas (a) and foils (b).

by a distinctive final thickness and, therefore, as the grain size decreases, the fraction of the material which is of the grain boundary nature increases. At the junctions of the grains there are partial disclinations (joint disclinations) in the chess order [36] of different power (or with different values of the Frank vector) and the fields of these disclinations partially compensate each other.

Of considerable importance for understanding the structure of the ultrafine-grained metals is the classification of the types of the boundaries in these metals. Insufficient attention has been paid to this problem by the researchers. At the same time, the metals (Al, Ni) subjected to severe plastic deformation by rolling have been classified by a detailed classification of the observed boundaries [37–39]. After all, not all the results of the authors of [37–39] are suitable for identifying the boundaries formed in metals in complex

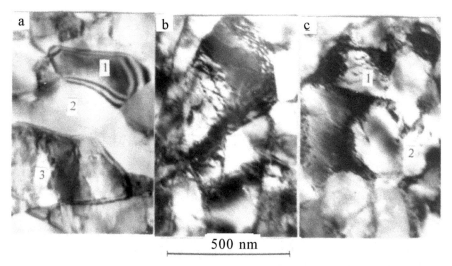

Fig. 4.7. Electron microscopic images of three types of grains and different boundaries in the ultrafine-grained nickel: a) the groups of grains (1, 2) without dislocations, delineated by the high-angle boundaries with the banded contrast and the grain (3) separated into fragments by high density dislocation walls; b) the groups of the grains containing the dislocations and delineated by the boundaries with the banded contrast; the boundaries contained dislocations; c – the groups of the grains (1, 2) containing dislocations and separated by high density dislocation walls.

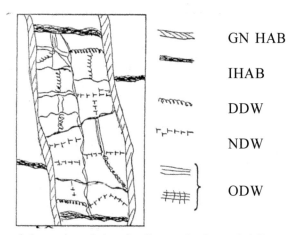

Fig. 4.8. The diagram of the grains of the ultrafine-grained metal delineated by the high-angle boundaries and containing different sub-boundaries: GN HAB – the geometrically necessary high-angle boundary; 4all; NDW – normal dislocation wall; ODW – the ordinary dislocation wall with banded contrast (with or without dislocations).

SPD treatments (ECAP, THP, etc.). Nevertheless, the experimental results show [14, 28] that the boundaries, observed in the ultrafine

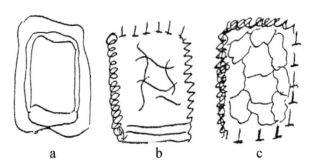

a   b   c

Fig. 4 .9. Diagrams of the grains delineated by different boundaries: a) the grains with no dislocations, b) the grain with a network dislocation structure, c) the grain with the cells and the fragments.

metals, are very similar to those systematised in [37–39]. This shows that when examining the boundaries in the ultrafine-grained materials it is sufficient to use the classification proposed by the authors in [37–39].

The electron microscopic images of the ultrafine-grained structure of Ni, observed on the replicas and foils, are shown in Figures 4.6 and 4.7. Detailed analysis of the similar images of the structure was used as a basis for constructing the scheme of the boundaries in this material (Figs. 4.8 and 4.9) [40]. The ultrafine-grained nickel contains the following types of boundaries. In particular, it is the geometrically necessary high-angle grain boundaries (GN HAB). The GN HAB are formed in the selected direction restricting several micrograins along their length. In Fig. 4.6, these boundaries are indicated by the arrows. In most cases, GN HAB are characterised by the banded contrast on electron microscopic images and do not contain lattice dislocations. They are imperfect in local sections. This is indicated by their only slightly distorted form. Evidently, in the course of formation of the ultrafine-grained structure considerable shear takes place at these boundaries. In the direction normal to the GN HAB there are incidental high-angle boundaries (IHAB). Together with GN HAB, they restrict the ultrafine grains. The IHAB are also characterised by the banded contrast but contain lattice dislocations. Evidently, these boundaries form by the storage of lattice dislocations in the previously formed boundaries of the subgrains. Finally, part of the ultrafine grains and, in particular, the largest of them, contain the cells or fragments (subgrains). The boundaries of the cells and fragments represent low-angle boundaries of different degrees of perfection (Fig. 4.9).

Second phase particles

It is important to mention another special feature of the structure of
the ultrafine-grained materials which is sometimes ignored by the
investigators. It is well-known that the ultrafine-grained materials
contain a high internal energy and their structure is characterised by
low stability. It can be stabilised by the nanosized particles of the
second phases distributed at the boundaries and in the junctions of
the grains and subgrains. In the two-phase ultrafine-grained materials
the second phases are produced intentionally [41, 42]. In the single-
phase, the so-called pure ultrafine-grained materials in which there
are always small quantities of the substitutional impurities, the second
phases form quite often in the course of severe plastic deformation
regardless of the purpose of the investigation [14, 43]. In fact, in the
course of severe plastic deformation the dislocations actively trap the
substitutional impurities and, in particular, the interstitial impurities,
and displace them to the boundaries. The large number of the point
defects and the high total density of other defects accelerate the
diffusion processes. The highly deformed metal additionally traps
the impurities from the surrounding atmosphere. Consequently, in the
course of severe plastic deformation, the ultrafine-grained metals are
characterised by the formation of the particles of the second phases
both equilibrium and metastable. These ultrafine particles, distributed
at the boundaries, in the junctions of the grains and subgrains and
at the dislocations, take part in the formation of the ultrafine-
grained structure. They increase the temperature threshold of the
recrystallisation of the structure. At the same time, the particles of
the second phases are sources of internal stresses. Detailed electron
microscopic studies of the specimens of the ultrafine-grained copper
and nickel, prepared by ECAP and THP, confirm the presence of
the nanosized particles of the second phases [14, 43]. Figure 4.10
shows the micrographs of bright and dark field images of some of
the observed particles in ultrafine-grained copper. Copper contained
particles of the phases Cu_3Sn (Sb), Cu_3N, Cu_2O and CuO. The first
phase forms from the impurities present initially in copper, the
remaining phases – from the nitrogen and oxygen trapped during
deformation. The ultrafine-grained nickel content particles of Ni_4N,
NiO and Ni_2O_3. The particle size varied in the range from 2 to 50 nm.
For example, the size of the Cu_3Sn (Sb) particles was approximately
50 nm, Cu_3N – from 10 to 40 nm, copper oxides – from 2 to 8 nm.
It may be seen that in the severe plastic deformation conditions

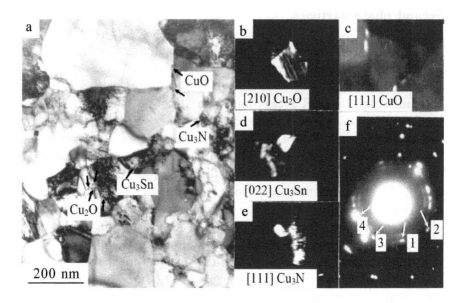

Fig. 4.10. Electron microscopic identification of the phases in ultrafine-grained copper, prepared by equal channel angular pressing: a) bright field image; b – e) the dark field images of the phases in the appropriate reflections; f) the microdiffraction patterns (I – [210] Cu_2O, II – [111] CuO, III – [022] Cu_3Sn, IV – [III] Cu_3N).

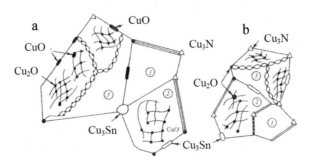

Fig. 4.11. Schematic pattern of three types of grains in the ultrafine-grained copper, prepared by ECAP (a) and THP (b), and the distribution of the particles in them: 1) the grains without the dislocations, 2) the grains containing dislocation networks, 3) the grains containing the cells and/or fragments.

the metastable phases form in most cases. The phase particles were localised at different defects, in particular, there were distributed at the grain boundaries and subboundaries and the joints. Figure 4.11 shows the distribution of the particles of the phases in the ultrafine-grained copper. It should be noted that the method of preparation of ultrafine-grained copper influences the volume fraction of the resultant particles: this fraction is higher in THP than in ECAP.

4.4. Sources of internal stress fields in ultrafine-grained materials

As mentioned previously, the ultrafine-grained materials are characterised by excess energy. This energy is localised in the defective structure of a large number of grain boundaries, the boundaries of the cells and fragments, at triple junctions of the grains, and in dislocation and disclination substructures. The large part of the excess energy of the ultrafine-grained material is the elastic energy of the distortion of the crystal lattice [44–46]. The previously mentioned crystal structure defects are sources of the field of internal stresses in the ultrafine-grained materials. The particles of the second phases, present in the ultrafine-grained metals, complicate the pattern of the field of the internal stresses even more.

In the experiments, detailed identification of the sources of the stress fields in the ultrafine-grained materials can be carried out by X-ray diffraction analysis because when using this method we determine the integral characteristic of the field of the internal stresses averaged out with respect to the sections of the specimens with the size of approximately several square micrometres. This problem can be solved by transmission electron microscopy. X-ray diffraction analysis provides in this case additional information for the internal stress fields.

The internal stress fields were studied by transmission electron microscopy in ultrafine-grained copper in [14, 27, 28, 43]. As shown previously, in electron microscopic studies they are detected on the basis of the bending extinction contours present on the images of the structure of the material. The typical micrographs for ultrafine-grained copper, demonstrating the existence of the extinction contours in the vicinity of different sources, are shown in Fig. 4.12. Analysis of the images shows that the dislocation structure is one of the sources of the internal stress fields. However, there are two main sources of the internal stresses – dislocation low-angle boundaries (Figure 4.12a–c) and the disclinations, localised at triple joints (Figure 4.12d, e). As the grain size decreases, the value of the Frank vector ω at the junctions of the grains increases. It is well-known that the Frank vector ω is proportional to the bending–torsion χ. There are two types of bending–torsion. In one case, the sources are dislocations whose field is screened by the dislocation structure

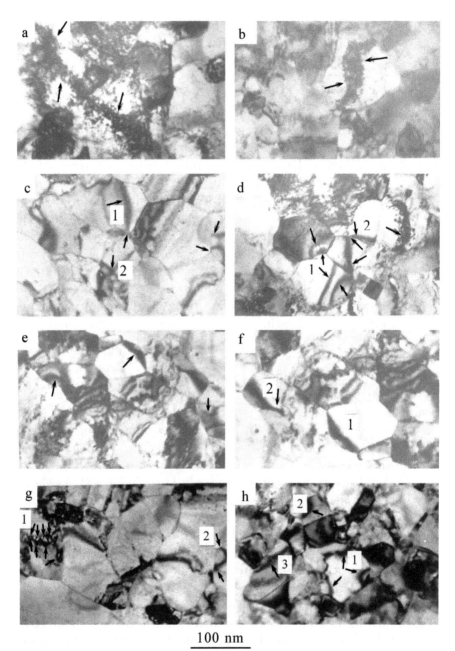

100 nm

Fig. 4.12. Examples of the microstructures with different sources of internal stresses; (a–b) – sub-boundaries; (c–e) – the joint dislocations; (f–j) – the particles inside the grains (g1, f2), the boundaries of the fragments (g2, d, j); (d, h) – the contact of the grains and high-angle boundaries (h2, h3); (k, l) – incompatibility of deformation of the adjacent grains. The numbers 1, 2, 3 the notes the different areas on the micrographs (for continuation of Fig. 4.12 see the following page).

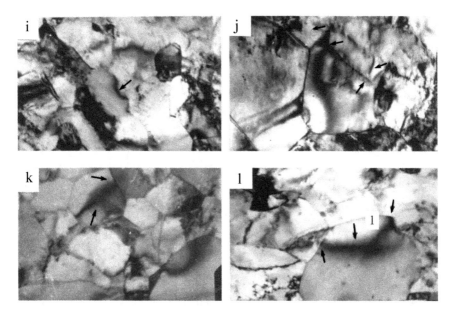

100 nm

Fig. 4.12 (continued.)

formed in the vicinity of this source. In another case, the field of the disclinations is not screened. Its amplitude is considerably greater than in the case in which screening takes place. The data for the variation of for the screened and non-screened cases of bending–torsion in dependence on the grain size in the conglomerate of the grains of UFG copper are presented in Fig. 4.13. The figure shows that the value of χ increases with a decrease of the grain size for both cases.

The grain boundaries of the ultrafine-grained materials are also sources of the internal stress fields (Figure 4.12k). However, the amplitude of bending–torsion from the grain boundaries is 1.5–2 times smaller than from the joint disclinations (Fig. 4.13). When the grain boundaries are the sources of internal stresses, as in the case of disclinations, there are non-screened fields of the internal stresses and the fields screened by the dislocations. With a decrease of the grain size and, correspondingly, increase of the degree of imperfection of the grain boundaries the bending–torsion from the grain boundaries increases (Fig. 4.13). Another source of the internal stress fields forms as a result of the incompatibility of deformation of the adjacent grains (Fig. 4.12 l).

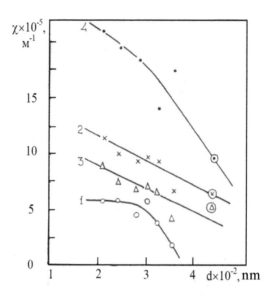

Fig. 4.13. Dependence of the curvature – torsion (χ) of the crystalline lattice on the grain size (d) in ultrafine-grained copper (1), (2) – the grain boundary sources of stresses; (3), (4) – the sources of the disclination type (1, 3 – screened bending – torsion; 2, 4 – non-screened bending-torsion; ⊙, ⊗, ○ – ECAP; ●, ◍ – THP.

The source of the internal stress fields are the particles of the second phases formed in ultrafine-grained copper during producing. In most cases, they are distributed at triple junctions of the grains (in the cores of the disclinations). Suitable examples are shown in Fig. 4.12d, g. The amplitude of the field from the second phase particles, distributed at the junctions of the grains, is of the same order of magnitude or even greater than in the case of the joint disclinations, and depends only slightly on the grain size. In ultrafine-grained copper it is mostly the particles of Cu_3N. Particles of CuO and Cu_2O were also found (Fig. 4.12c, g). These particles are small (~5 nm). They are often distributed in the volume of the grains. In most cases, there are no screening dislocations around these fine particles. The value of χ in this case is high and changes with a decrease of the grain size in the range $(10–50) \cdot 10^5$ m^{-1}.

It should be noted that the internal stress fields, produced by different sources, may interact. Consequently, the fields overlap and the overall pattern of the internal stress fields becomes more complicated. This is clearly indicated by the sections of the images of the structure where different ends of the bending extinction contours close on different sources (Fig. 4.12c, d, f, g, j, k).

Table 4.1. Amplitude (τe) of the field of internal stresses in the vicinity of various sources in ultrafine-grained nickel (transmission electron microscopy and X-ray diffraction analysis data)

Area in which stress fields are measured	τ_e, MPa ($\pm$50 MPa)	
	ECAP	ECAP + annealing (125°, 3 hours
Grain and subgrain boundaries	255	240
Subgrain boundaries	260	250
Cell boundaries	350	410
Banded grain boundaries with dislocations	130	210
Dense dislocation walls	230	240
Inside grains of different type	170	120
Inside grains with network dislocation structure	160	100
Inside dislocation-free grains	130	130
Mean value of τ_e from different sources (TEM)	247	215
Mean value in X-ray diffraction analysis	232	192

The long-range fields, produced by the disclinations, the grain boundaries and the particles, encircle grains of different type. The amplitude of the elastic field χ from the sources in all types of grains usually increases with a decrease of the grain size. The dependence of the value of χ on the type of grains is considerably weaker. It is maximum in the dislocation-free grains and slowly decreases in transition to the grains with the dislocations and then to the grains with fragments. An important mechanism of the screening of the elastic fields of disclinations and particles – the redistribution of the dislocations, in particular the presence of a correlation (ordering) in the dislocation ensemble [15, 47, 48]. Purely disclination relaxation mechanisms and mechanism of screening the elastic fields of the internal stresses can also operate [49].

The sources of the internal stress fields, detected in copper, are also typical of ultrafine-grained nickel [27]. The experimental results show that, as in ultrafine-grained copper, the most powerful sources of the internal stress fields in ultrafine-grained nickel are the low-angle boundaries, the junctions of the grains and the grain boundaries. These follows in particular from the data in Table 4.1. The table gives the mean values of τ_e, measured on the basis of the curvature of the dislocation lines in different local sections of the ultrafine-grained nickel in different states (in the initial state, after severe plastic deformation, and after low-temperature annealing). Table 4.1 shows the mean values of the stresses in the specimens measured by X-ray diffraction analysis. Comparison of these values with the values of τ_e averaged out with respect to all sources (obtained by the transmission electron microscopy) shows that, firstly, there is a good agreement of the values of τ_e measured by the individual methods. Secondly, since the X-ray diffraction analysis provides only the average value of τ_e in the thick specimens, then consequently transmission electron microscopy makes it possible to present the actual pattern of the field of the internal stresses in the ultrafine-grained material for analysis of these data confirm is that the boundaries of the dislocation cells are characterised by higher stresses and the grain boundaries and the subgrain boundaries.

As indicated by Table 4.1, low-temperature annealing changes only slightly the overall pattern of the internal stresses. Nevertheless, Table 4.1 shows that annealing often increases the values of τ_e. This indicates that the internal stress fields controls the evolution of the defective structure of the ultrafine-grained material. In fact, the value of τ_e, determined by the boundaries of the cells, increases after annealing ultrafine-grained nickel. This indicates that the defective structure is rearranged during annealing. New boundaries being the sources of even higher stresses them prior to annealing, appear in the structure. The internal stresses, generated by the disclinations, distributed in the triple junctions of the ultrafine-grained nickel, were also studied using the bending extinction contours. The results show that there are joint disclinations in 20% of triple junctions in the initial state of ultrafine-grained nickel. The Frank vector (ω) is quite high: the average value of χ reaches $1 \cdot 10^6$ m^{-1}. Annealing slightly decreases the number of these junctions and also the value of χ. The magnitude of the stresses is the highest in the vicinity of the source of the internal stresses and decreases with increase of the distance from the source.

The above described spectrum of different sources of the internal stress fields in the investigated ultrafine-grained copper and nickel is in agreement to a large extent with the models of the structure of the ultrafine-grained material, presented in [50–52]. The internal stress fields in the sources also include the joint disclinations, and the grain boundaries. The experiments show [14, 43] that the models do not take into account another source of internal stresses – the second

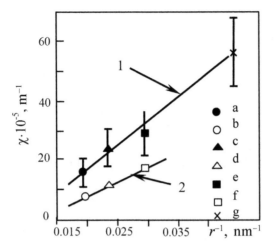

Fig. 4.14. Dependence of the bending–torsion (χ) of the crystalline lattice in the vicinity of the non-deformed particles for the non-screened (1) and screened (2) bending–torsion on the particle size (r): (a, b – the particles of Cu_3Sn; c, d – CuO; e, f – Cu_3N; g – CuO and Cu_2O). In the case (a–f) the particles are distributed at triple junctions and the grain boundaries, in the case (h) – inside the grains.

phase particles. The curvature–torsion of the crystal lattice, formed locally in the vicinity of the second phase particles, depends on the size of these particles. The magnitude of distortion of the crystal lattice increases with a decrease of the particle size. Theoretical considerations regarding the stress concentration in the elastic matrix in the vicinity of the non-deformed particles predict the dependence $\chi \sim r^{-1}$ [53]. Study [43] presents the average data for χ and r for the cases in which the non-screened bending–torsion of the crystal lattice, formed as a result of the presence of the particles in the ultrafine-grained copper. The appropriate results are presented in Fig. 4.14 (curve 1). In fact, the dependence of χ on r^{-1} is linear. This means that the elastic approximation can be used in the theory in the plastically deformed matrix in dislocation-free areas. It is curious that in the case of the bending–torsion screened by the dislocations the linear dependence of χ on r^{-1} remains unchanged (Fig. 4.14, curve 2).

The sources of the internal stresses in the ultrafine-grained materials, prepared by severe plastic deformation, are almost of the same nature as in the deformed polycrystals with the conventional grain size [17]. It is the dislocation structure, the boundaries of the grains and subgrains, their junctions, the incompatibility of the deformation of the adjacent grains, and the second phase particles.

However, the magnitude of the internal stresses in the ultrafine-grained metals is higher (by approximately an order of magnitude) [17]. The spectrum of the sources of the internal stress fields in the ultrafine-grained metals does not contain dislocation pile ups and broken boundaries, present in the deformed polycrystal with the conventional grain size [17]. It should be mentioned that in the ultrafine-grained materials because of the large length of the boundaries, different types of boundaries and a large number of the triple junctions, the role of the internal stresses in the processes taking place in these materials is after all far more significant than in the polycrystals with the conventional grain size.

The electron microscopic measurements of the internal stresses in the grains of the ultrafine-grained materials make it possible to solve an important problem discussed previously. This problem is the determination of the actual grain size. X-ray diffraction analysis is used for these purposes, in addition to transmission electron microscopy. The determination of the true grain size in the ultrafine-grained materials is not a simple task. As already mentioned, the interpretation of the transmission electron micro-scopic data is complicated because of the complicated contrast of different boundaries and sub-boundaries. The results obtained by X-ray diffraction analysis are strongly affected by the internal stress fields. Therefore, it is interesting to compare the dimensions of the substructural formations, determined by transmission electron microscopy and X-ray diffraction analysis. In [28] these data were obtained for ultrafine-grained nickel. In X-ray diffraction analysis, the size of the substructural formations was determined from the half width of the lines after separating the contribution of microdistortions. Figure 4.15 shows the comparison of the transmission electron microscopy and X-ray diffraction analysis data. After taking into account all the corrections, it may be concluded that the so-called 'X-ray' size of the micrograins is the average size of the subgrains (fragments).

4.5. Distribution of internal stresses in grains. The scheme of the grains of ultrafine-grained materials

At present, there is a number of models of the grains of ultrafine-grained materials [25, 33–35]. These models consider a hardened boundary zone. Examination of the distribution of the internal stresses by transmission electron microscopy makes it possible to detect this

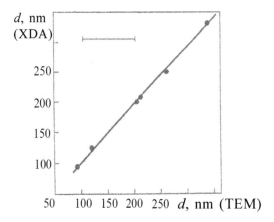

Fig. 4.15. The correlation between the size of the subgrains, measured by transmission electron microscopy (TEM), and coherent scattering blocks, measured by X-ray diffraction analysis (XDA). Ultrafine-grained nickel, prepared by equal channel angular pressing.

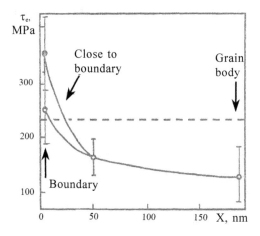

Fig. 4.16. Variation of the internal stresses (τ_e) with distance from the boundaries of the cells ($\bullet$) and grains (o). UFG nickel prepared by equal channel angular pressing.

zone in the grains by experiments (Fig. 4.16). It is interesting to determine the distribution of the internal stresses in the volume of the grain. Figure 4.17 shows the schematic typical distribution of the internal stresses for two types of grains of ultrafine-grained copper: 1) containing dislocation networks (dislocation chaos), and 2) containing the boundaries of the cells and fragments. The internal

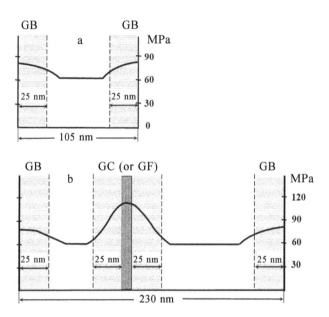

Fig. 4.17. Distribution of internal stresses (τ_e) (vertical axes) in the grains in ultrafine-grained copper, produced by equal channel angular pressing: a – for the grains containing the chaotically distributed dislocations and dislocation networks; $d = 105$ nm; b – for the grains containing dislocation cells and fragments, $d = 230$ nm. GB is the grain boundary, GC and GF are the boundaries of the cells and fragments, respectively. Crosshatching indicates the hardened regions in the vicinity of the grain boundaries and the boundaries of the cells and fragments.

stresses were determined on the basis of the data for the curvature of the free dislocation segments. Figure 4.17 shows that in the grains of both types in the vicinity of the grain boundaries there are hardened zones with a higher level of the stresses in comparison with the body of the grain. The average width of the zones in both types of grains is approximately 25–35 nm. In the body of the grains, containing dislocation networks, the average level of the stresses is $\tau_e = (64\pm35)$ MPa (Fig. 4.17a). The internal stresses in the vicinity of the grain boundaries increase to (80 ± 90) MPa. The high dispersion of the values of τ_e (35 MPa and 90 MPa) is determined by the presence of different sources of the internal stresses at the grain boundaries. The pattern in the grains containing dislocation cells or fragments slightly differs (Fig 4.17b). The difference is the presence inside the grains of the low-angle boundaries of the cells and fragments. It may be seen that these are the regions of the grains of ultrafine-grained copper subjected to the highest stresses. Here the internal stresses reach (110 ± 130) MPa. The highly stressed boundaries of the cells and fragments relax in the course of severe plastic deformation, and transform to the high-angle grain boundaries. This results in the rearrangement of the grain structure because the cells and fragments are replaced by new grains with the chaotic or network dislocation structure. The investigations show that the distribution of the internal stresses in the grains during deformation of the ultrafine-grained copper does not change greatly.

The description of the grains of the ultrafine-grained materials, represented here on the basis of the experimental measurements of the internal stresses, indicates the possibility of using the composite model for examining the work hardening of these materials. At the moment, there are several variants of the composite models of hardening of the polycrystals with the conventional grain size. They always contain the hardened zone. In the Westbrook–Arkharov model [54] it is the boundary layer with a higher concentration of the impurities in comparison with the body of the grain. In the Kocks–Hirth model [55] it is postulated that the grain with the size of d is in fact a composite. The grain is divided into two regions: $d = dx + dy$, where dx is the hardened boundary zone with the deformation resistance τ_x, differing from the resistance τ_y of the internal part of the grain. These assumptions were developed further by H. Mughrabi [5, 56] for dislocation cells. In this model, the boundary of the cell and boundary zone are combined into a wall with a high dislocation density. The internal stresses in the wall are higher than inside the cell, and this is confirmed by X-ray diffraction analysis [5].

The authors of [57, 58] determined by experiments the differences in the dislocation structure of the body of the grain and the boundary zone in the polycrystals with the conventional grain size. In the boundary zone of the grains the dislocation density is higher, the grain and fragment sizes are smaller, the extent of curvature–torsion of the crystal lattice and the internal stresses are higher. In the model of the grains of the polycrystal, proposed in [40], all these factors are taken into account.

Recently, because of the special attention paid to the investigations of the ultrafine-grained materials, various investigators have proposed composite models for the fine-grained polycrystals. They develop the concepts previously proposed for coarse-grained materials. In the grain model proposed by Kim–Estrin–Bush [35] there are grain boundaries at which sliding can take place and triple junctions, quadripolar junctions and the body of the grains can take place. The composite nature of the model is ensured by the separation of these four regions of the grain with different mechanical characteristics.

The model proposed by R.Z. Valiev et al [25, 26] is based on the results of X-ray diffraction analysis and the data for the elastic fields of the internal stresses calculated for different dislocation-disclination configurations in the ultrafine-grained materials. The model is based on the following special features of the structure of the grain boundaries in the boundary zone: the defective structure of the non-equilibrium grain boundaries and the boundary zone, the presence of disclinations at the junctions of the grains. The thickness of the hardened boundary zone according to the estimates provided by R.Z. Valiev et al is 10–20 nm which is in good agreement with the results of the experimental studies carried out by the authors of the present book (Fig. 4.17).

The above results show that the scheme of the grains of the ultrafine-grained materials, proposed in the book, is in satisfactory correlation with composite models. The experimental results indicate that the boundaries and near-boundary regions of the dislocation cells and fragments (subgrains) are the areas of the ultrafine-grained materials subjected to the highest stresses. This is followed, in the order of decreasing internal stresses, by the junctions of the grains (joint disclinations), grain boundaries and second phase particles.

4.6. Conclusions

The experimental results show the existence of elastic long-range fields of the internal stresses in the ultrafine-grained metallic materials. The high defectiveness of the ultrafine-grained materials, produced by severe plastic deformation, ensures the more important role of the internal stresses in the formation of the properties in comparison with the polycrystal with the conventional grain size. The internal stress fields are determined by the complicated structure of the ultrafine-grained materials, containing different types of grains with different defectiveness, and a large variety of the boundaries with different degrees of perfection.

Transmission electron microscopy is a method of providing a large amount of information for examining and measuring the internal stresses. The method has been used for detailed examination of the internal stress fields in the ultrafine-grained nickel and copper. The experimental results show that the spectrum of the sources in the ultrafine-grained metals is very similar to that detected in the polycrystal with the conventional grain sizes. It is the dislocation structure, the grain boundaries, the boundaries of the subgrains and cells, the junctions of the grains, and the second phase particles. There are typical differences in the spectrum of the sources in the ultrafine-grained metals in comparison with the conventional polycrystals. The ultrafine-grained materials do not contain dislocation pile ups and broken boundaries. These are often present in the conventional deformed polycrystals and also act as the sources of the internal stress fields.

An important property of the ultrafine-grained materials is a high level of the internal stresses (by an order of magnitude or more) in comparison with the conventional deformed polycrystals. The source of the highest internal stresses in the ultrafine-grained materials are the boundaries of the dislocation cells and fragments, disclinations at triple junctions and the second phase particles. The relaxation of the stresses in the process of severe plastic deformation is accompanied by the formation of high-angle boundaries and, correspondingly, the formation of new micrograins.

The results of experimental studies of the internal stresses by transmission electron microscopy have been used to present a scheme of the distribution of the internal stresses in grains of different type. According to this scheme, the areas in the vicinity of the grain boundaries and the boundaries of the cells and fragments contain the zones with the highest internal stresses. The level of the internal stresses increases with a decrease of the grain size and this is caused by the following reasons. As the grain size decreases, the efficiency of screening of the elastic fields of the internal stresses of the dislocations decreases as a result of: 1) decrease of the distance to the source of the stresses, and 2) decrease of the dislocation density capable of screening the fields of the internal stresses. The hardened zones in the vicinity of the boundaries are found in a number of theoretical models of the ultrafine-grained materials. Similar zones were also detected in experiments in polycrystals with the conventional grain size. However, the fraction of the volume of the grain in the vicinity of the grain boundaries and with the higher level of the internal stresses in comparison with the body and the ultrafine-grained material is greater.

In transition from the coarse-grained polycrystals to the ultrafine-grained materials the nature of the hardened grains of changes: the role of the contribution to the hardening of the elastic fields from the grain boundaries and the junctions of the grains in these zones increases. In the polycrystals with the conventional grain size the main mechanism of work hardening of the zone is the higher density of the defects in the vicinity of the grain boundaries. In the ultrafine-grained materials, this is accompanied by high internal stresses.

References

1. Seeger A., Dislocations and Mechanical Properties of Crystals, New York, J. Wiley and Sons, 1957. P. 243 – 329.
2. Mughrabi H., *Acta Metall.*, 1983, V. 31, No. 9, 1367–1379.
3. Ungar T., et al., *Acta Metall. Mater.*, 1984, V. 32, No 3, 333–342.
4. Koneva N.A., et al., Theoretical and experimental study of disclinations, Leningrad, A.F. Ioffe Institute, 1984, 161–167.
5. Mughrabi H., *Mat. Sci. Eng.,* 1987, V. 85, 15–31.
6. Koneva N.A., et al., Strength of Metals and Alloys. Proceed. ICSMA–8, Oxford: Pergamon Press, 1988, 385–390.
7. Muller M., et al., *Z. Metallkd.*, 1995, V. 86, No. 21, 827–831.
8. Argon F.S., Haasen P., *Acta Metall. Mater.*, 1993, V. 41, No. 11, 3289–3306.
9. Gunter B., et al., *Phil. Mag.* B, 1993, V. 68, 825–832.
10. Warren B.E., X-ray Diffraction, Reading, Menlo Park, London. Don Mills: Addison-Wesley Publishing Company, 1968.
11. Lipson G., Steeple G., Interpretation of the powder X-ray patterns, Moscow, Mir, 1972.
12. Straub S., et al., *Acta Mater.*, 1996, V. 44, No. 11, 4337–4350.
13. Ungar T., *Mater. Sci. and Eng.*, 2001, V. A309–310, 14–22.
14. Kozlov E.V., et.al., *Ann. Chim. Fr.*, 1996, V. 21, 427–442.
15. Kozlov E.V., Koneva N.A., *Mat. Sci. Eng.*, 1997, V. A234–236, 982–985.
16. Mughrabi H., *Mater. Sci. and Eng.*, 2001, V. A309–310, 14–22.
17. Koneva N.A., et al., *Mat. Sci. Eng.*, 2001, V. A319–321, 156–159.
18. Tyumentsev A.N., et al., Theoretical and experimental study of disclinations. (Ed. V.I. Vladimirov), Leningrad, A.F. Ioffe Institute, 1984, 173–180.
19. Koneva N.A., et al., *Izv. AN Ser. Fizicheskaya*, 1998, V. 62, No.7, 1352–1358.
20. Konstantinova T.E., Mesostructure of deformed alloys, Donetsk, DonFTI, 1997.
21. Hirsh P., et al., Electron microscopy of fine crystals, Moscow, Mir, 1968.
22. Koneva N.A.,et al., New physical and mechanical methods of research of materials in loading (ed. V.E. Panin), Tomsk, TSU, 1990, 83–93.
23. Strunin B.M. *Fiz. Tverd. Tela,* 1967, V. 9, No. 3, 805–812.
24. Segal V.M., et al., Processes of plastic structure formation of metals, Minsk, Nauka i tekhnika.
25. Valiev R.Z., Alexandrov I.V., Nanostructured materials, produced by severe plastic deformation, Moscow, Logos, 2000.
26. Valiev R.Z., et al., *Progr. Mater. Sci.*, 2000, V. 45, 1–103.
27. Koneva N.A., . et.al. , Investigations and Applications of Severe Plastic Deformation, New York, Kluwer Acad. Publ, 2000, 121–126.
28. Kozlov E.V., et. al, Ultrafine-grained Materials II. Warrendale: A publication of the Minerals, Metals and Materials Society, 2002, 419–428.
29. Ivahashi Y., et al., *Acta Metall. Mater.*, 1998, V. 46, No. 9, 3317–3331.

30. Koneva N.A., et.al., Evolution of the Structure of Ultrafine-grained Materials III. Warrendale: The Minerals, Metals and Materials Society, 2004, 397–402.
31. Kozlov E.V., et al., *Metally,* 1993, No. 5, 152–161.
32. Abdulov R.Z., et al., *J. Mat. Sci. Letters,* 1990, V. 9, 1445–1447.
33. Valiev R.Z., *Nanostructural Materials,* 1995, V. 6, No. 1–4, 73–82.
34. Valiev R.Z., Alexandrov I.V., *Nanostructural Materials,* 1999, V. 12, 35–40.
35. Kim H.S., et al., *Mater. Sci. Eng.,* 2001, V. A316, 195–199.
36. Rybin V.V., *Voprosy materialovedeniya,* 2002, V. 1, No. 29, 11–33.
37. Bay B., et al., *Mat. Sci. Eng.,* 1989, V. A113, 385–397.
38. Hughes D.A., Hansen N., *Acta Metall. Mater.,* 2000, V. 48, 2985–3004.
39. Wert S.A., et al., *Acta Metall. Mater.,* 1995, V. 43, No. 11, 4153–4163.
40. Koneva N.A., et.al., Ultrafine Grained Materials, II. Warrendale: A publication of the Minerals, Metals and Materials Society, 2002, 505–514.
41. Morris D.G., Morris M.A., *Acta Met.,* 1991, V. 39, No. 8, 1763–1770.
42. Apps P.J., et al., *Acta Metal. Mater.,* 2003, V. 51, 2811–2822.
43. Koneva N.A., et al., Structure and phase transformations and properties of nanocrystalline alloys (Eds. G.G. Taluts and N.I. Noskova), Ekaterinburg, Ural Branch of Russian Academy of Sciences, 1997, 125–140.
44. Tschope A., Birringer R., *Acta Met. Mater.,* 1993, V. 41, 2791–2796.
45. Tschope A., Birringer R., Gleiter H., *J. Appl. Phys.,* 1992, V. 71, No. 11, 5391–5394.
46. Gertsman V.Y., et al., *Phys. Stat. Sol.* (A), 1995,V. 149, 243–252.
47. Koneva N.A., Kozlov E.V., *Izv. AN Ser. Fiz.,* 2002, V. 66, No. 6, 824–821.
48. Koneva N.A., et al., *ibid,* 2002, V. 68. No. 5, 638–640.
49. Gryaznov V.G., Trusov L.I., *Progr. in Mater. Sci.,* 1993, V. 37, 289–401.
50. Valiev R.Z., *Nanostructural Materials,* 1995, V. 6, No. 1–4, 73–82.
51. Romanov A.E., *Nanostructural Materials,* 1995, V. 6, 125–134.
52. Nazarov A.A., et al., Nanostructural Materials, 1993, V. 4, No. 1, 93–101.
53. Aleksandrov V.M., et al., Thin stress concentrators in elastic bodies, Moscow, Fizmatlit, 1993.
54. Gleiter G., Chalmers B., Large-angle grain boundaries Moscow, Mir, 1975.
55. Hirth J.P., *Metal. Trans.,* 1972, V. 3, 3047–3067.
56. Mughrabi H., et al., *Phil. Mag.* A, 1986, V. 53, No. 6, 793–813.
57. Perevalova O.B., Koneva N.A., *Fiz. Met. Metalloved.,* 2003, V. 95, No. 4, 106–112.
58. Perevalova O.B., et al., *Fiz. Met. Metalloved.,* 2003, V. 95, No. 6, 85–93.

5

Severe plastic deformation

5.1. Introduction

Extreme effects exert a strong influence on the structure and properties of solids [1]. Undoubtedly, they include the effect of severe plastic deformation. Recently, there has been special interest in this method of effect on the materials because it makes it possible to produce special nanostructured states and, at the same time, increase greatly the physical and mechanical properties of metallic materials [2]. As a result of the effect of Soviet (Russian, Ukrainian and Belarussian) scientists, the interest in severe plastic deformation has reached the worldwide scale. Starting with the First International conference in Moscow in 1999, regular international conferences NanoSPD concerned with this problem have been organised throughout the world.

The structural states formed under high strains are highly unusual and difficult to predict because the classic disclination and dislocation approaches, used efficiently for describing conventional deformation, are insufficiently effective in this case. Unfortunately, most of the investigators limit their attention to the examination of final structures and the appropriate properties of the materials after different conditions of high plastic deformation without analysing the nature of the physical processes which take place directly under extremely high degrees of plastic flow. Nevertheless, recently there has been a large number of important fundamental studies which can be used to a certain extent to explain the physical nature of the structural processes typical of high plastic deformation. The subject of this chapter is a generalised analysis of these investigations.

Studies of the physics of severe plastic deformation started undoubtedly by studies of the Nobel Laureate P.W. Bridgman who for many years carried out experiments to study severe plastic deformation and fracture of crystalline (metals and alloys) and amorphous (plastics and boron anhydride B_2O_3) solids under high pressure in tensile, compression, torsion, and shear tests and other types of the stress state [3]. The original systems constructed by Bridgman were used to detect for the first time new phenomena taking place under high deformation. It is important to mention the brave hypothesis proposed by Bridgman: ''it is possible that the crystal (under high deformation) breaks down into a mosaic of smaller and smaller crystals, and the structure inside crystals changes only slightly...'' [3]. In fact, Bridgman, not knowing the theory of dislocations, predicted the fragmentation phenomenon which, as explained later, plays the fundamental role under high deformation. On the basis of the experiment he also predicted the possibility of transition under high deformation from the crystalline to amorphous state: ''at very high shear strains the crystal structure of copper fractures to such an extent that only one greatly diffuse line can be produced on the Debye pattern. Evidently, this means that the linear dimensions of the grains are smaller than 10 Å (1–2 nm)'' [4]. Although Bridgman used equipment with different types of stress state, at present only the Bridgman chamber (Fig. 5–1) is used in which SPD at high quasi-hydrostatic pressure (several GPa) is implemented by torsion with the variation of the angle of rotation of a mobile anvil in relation to a stationary one [5].

A further impetus in the investigations of severe plastic deformation is associated with the studies of V.M. Segal [6] and R.Z. Valiev [7] who proposed to generate SPD by the method of equal

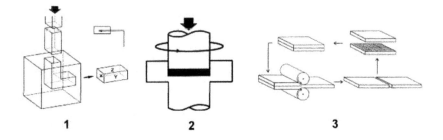

Fig. 5.1. Diagram of the most widely encountered methods of producing severe plastic deformation: 1 – equal channel angular pressing, 2 – torsion under pressure in the Bridgman chamber, 3 – accumulated rolling.

channel angular pressing (ECAP) (Fig. 5–1). Later, special structural states, formed under severe plastic deformation, were produced by the methods of screw extrusion, accumulating rolling (Fig. 5–3), free forging and other methods [8].

The plastic deformation formed in these methods is so high that the conventional values of the relative degrees of deformation lose their meaning and it is necessary to transfer to the true logarithmic strains e. Their values are determined the following form.

For deformation in the Bridgman chamber [4]:

$$e = \ln\left(1+\left(\frac{\varphi \cdot r}{h}\right)^2\right)^{0.5} + \ln\left(\frac{h_0}{h}\right), \qquad (5.1)$$

where r and h are respectively the radius and height of the disc-shaped specimens, processed in the Bridgman chamber, and φ is the angle of rotation of the moving anvil. The number of the total rotations of the moving anvil N corresponds to the strain at which $\varphi = 2\,\pi N$.

For equal channel angular pressing [5]:

$$e = \operatorname{arsh}\left(n \cdot \operatorname{ctg} \phi\right) \qquad (5.2)$$

where n is the number of passes and ϕ is the angle of rotation of the channels.

5.2. Terminology

Taking into account the assumptions of the pioneers in the area of examination of superhigh strains, a similar process in which the true strain e is higher than 1 and may reach the values of 7–8 or greater, this type of deformation is referred to as severe plastic deformation. However, this term is not completely accurate. In fact, intensive processes in nature are usually regarded as the processes taking place at a high rate [9]. At high deformation, as shown by the estimates, the strain rate is in the range 10^{-1}–$10^{1}\,\text{s}^{-1}$, i.e., in the transition range between the static and dynamic strain rates, corresponding to the rates obtained, for example, in conventional rolling. Therefore, this type of deformation cannot be regarded as intensive. In addition, as noted for the first time by M.V. Degtyarev [10], from the physical viewpoint the process of plastic deformation should not be regarded as an intensive but as extensive process. The term severe plastic deformation used in the foreign literature [1] is evidently more suitable.

The authors of [13] proposed another, physically more suitable Russian language term which is applicable to the general concept of our considerations regarding the surrounding matter. It is well-known [14] that natural science examines three scale levels of the material world: *the microworld* in which the scale of the individual atoms and molecules is observed, *the macroworld* – the scale of the human perception of the world (meter, kilogram, second), and the *megaworld* – the astronomical scale. There is a direct analogy between the previously described scale levels of organisation of matter and the levels of plastic deformation. In fact, we know the process of *microplastic* deformation [15] observed up to reaching the value of the macroscopic yield limit, and the process of *macroplastic* deformation,observed at stresses above the yield limit [9]. Thus, continuing this analogy, the region of severe plastic deformation should be referred to as the *severe plastic* deformation (SPD) which corresponds to the general logic of development of any material phenomenon.

If the boundary between the microplastic deformation and the macroplastic deformation is determined quite accurately – the degree of deformation corresponding to the macroscopic yield limit (relative strain $\varepsilon = 0.05$ or 0.2%), the boundary between the microplastic deformation and severe plastic deformation has not as yet been determined. Conventionally, the boundary region will be represented by the relative strain $\varepsilon \approx 100\%$ or the true strain $e \approx 1$. Further, a stricter, physically justified value of the plastic strain, corresponding to the transition to the SPD range will be formulated.

5.3. Structural models

The structural states formed under such gigantic deformation are highly unusual and difficult to predict. Unfortunately, the large majority of the authors, studying the effect of the severe plastic deformation have restricted that attention to the examination of the final structures and the corresponding properties of the materials, without analysing the physical processes which take place directly under gigantic degrees of plastic flow. The classic dislocation and also disclination approaches to understanding the structural processes under severe plastic deformation appear to be insufficiently effective and require re-examination.

Summing up the large number of experimental studies carried out to investigate the structure of the materials subjected to SPD, one can define the following processes and phenomena typical exclusively of the SPD [16]:

1. Fragmentation. Formation of a large fraction of high-angle grain boundaries.

2. Absence of work hardening.

3. The processes of low-temperature dynamic recrystallisation.

4. Anomalously high diffusion mobility of the atoms.

5. Active phase transformation processes, accompanied by the precipitation and/or dissolution of metastable and equilibrium phases.

6. Amorphisation processes.

7. Cyclic nature of the structural and phase transformations.

Unfortunately, there is no strict theory of structure formation during SPD in the literature which would be capable to explain from the united physical viewpoint the existence of all these phenomena. In addition, there are no distinctive views regarding the conditions of transition of plastic deformation from conventional macroscopic deformation to severe plastic deformation [17]. In addition, in order to examine the SPD as a structure-determined specific stage of plastic deformation, typical of all solids (Fig. 5.2), many investigators assume incorrectly that SPD is a special type of plastic deformation which can take place only under specific types of stress state in the pure shear conditions [2].

A details systematisation of the defective structures, formed in different materials with increase of the degree of plastic deformation, was carried out by N.A. Koneva and E.V. Kozlov [18]. They show that when approaching the SPD region, the structural states change to others (cellular, banded, fragmented structures, etc) similar to the structural phase transitions, depending on the nature of the material. This is accompanied by the change of the internal stresses in the

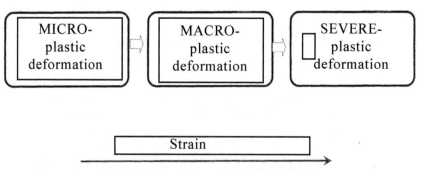

Fig. 5.2. Main stages of plastic deformation.

conditions for the occurrence of the anomalies of the mechanical behaviour of the crystal. In [19] it is shown theoretically that with increase of the SPD the structure shows the formation of a very large number of excess point defects (mostly vacancies) capable of stimulating diffusion phase transformations during deformation.

Special attention has been paid to the hypothesis according to which the SPD results in the formation of the 'unique' non-equilibrium grain boundaries [20]. These boundaries, according to many authors, are responsible for the anomalous phenomena of sliding, diffusion, interaction with lattice defects and, consequently, may also be responsible for the high values of strength and ductility.

It would be important to pay the attention to the results of unjustifiably forgotten and not very often cited study by V.A. Likhachev et al [21] in which a copper wire was subjected to SPD (e = 1.6 and 3.7). The authors detected cyclic changes of the structure with increasing strain: the fragmented structure (d = 0.2 μm) → the recrystallised structure → the fragmented structure (d = 0.1 μm), where d is the average fragments size. It is interesting to note that the fragmented structure of the 'second generation' is twice as fine as the structure of the 'first generation' (Fig. 4.5). In principle, this suggests the establishment of the nanostructured state in the third and further cycles.

The most elegant concept of fragmentation under high plastic deformation has been proposed by V.V. Rybin [22] and his colleagues [23]. On the basis of the assumptions of the dominant role of the disclination mode in high plastic deformation and the associated processes of fragmentation, Rybin described correctly the phenomena taking place at high strains, close to e = 1. Figure 5.3 shows the electron microscopic image of a disclination dipole in this stage of formation of the fragmented structure in cold rolling of molybdenum. In accordance with the disclination concept the size of the fragments – the main structural elements – smoothly decreases with increasing strain, reaching a constant minimum value of the order of 0.2 μm and then does not decrease any further.

The stage of active fragmentation was determined as the *developed plastic deformation*. In reality, this means that the transition to the region of the nanostructured state and the formation of the fragments (grains) smaller than 100 nm (0.1 μm) under the effect of the disclination mode in this stage of developed deformation is hardly possible. Analysis of a large number of experimental data, especially for the BCC crystals, made V.V. Rybin to assume the formation of

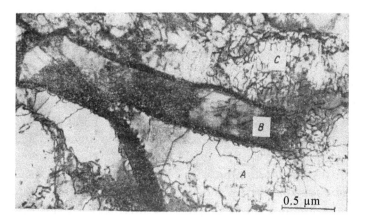

Fig. 5.3. Dipole configuration (B) made of two broken inclined boundaries in molybdenum in the stage of developed plastic deformation [22].

the limiting (critical) fragmented structure with the further evolution of the structure in the disclination mode no longer possible [22]. The regions of fracture form at the boundaries of the fragments which separate regions usually free from dislocations (Fig. 5.4). According to the author, the critical fragmented structure is the final product of plastic deformation, it is not capable of resisting the increasing effect of the external and internal stresses and should result in fracture.

The previously described critical fragmented structures formed at $e \leq 2$ for the case of uniaxial tensile loading or rolling of the BCC metals. The following question should be answered: would it be possible to obtain in the SPD stage in the process of a large number of experiments considerably higher strains $e > 1$–2, up to 10?. Unfortunately, the theory of developed plastic deformation does not provide answer to the question why it is followed by the SPD stage.

5.4. Energy principles of the mechanical effect on the solid

The fully efficient theory of the SPD should, according to the authors of the present book, be capable of providing an unambiguous answer to the following questions:

– what are the reasons for the special structural and phase transformations taking place during SPD?;

– what are the prerequisites for the occurrence of the SPD in accordance with a specific scenario?;

– what are the conditions of formation of the true nanostructured state in the SPD?

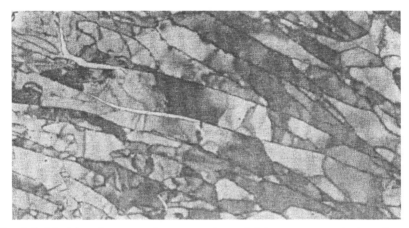

Fig. 5.4. Limiting fragmented structure in molybdenum; transmission electron microscopy, magnification ×40 000.

– what is the difference between SPD and 'conventional' plastic deformation?

– what determines the boundary value of deformation, starting at which it can assumed that SPD is taking place?

Attention will be given to the behaviour of a solid under loading from the viewpoint of a mechanical dissipative system in which the total deformation energy continuously decreases or is scattered, transferring to other non-mechanical forms of energy (elastic, chemical, electromagnetic, thermal, etc (Fig. 5.5) [24]. In other words, in the mechanical effect on the solid of finite dimensions a specific amount of energy is 'pumped' into the solid. The evident 'dissipation channel' of elastic energy is plastic deformation. When this channel becomes exhausted, the alternative channel may operate – mechanical fracture. However, at high values of the elastic energy plastic deformation can in principle initiate the additional 'dissipation channels': dynamic recrystallisation, phase transformations and additional release of thermal energy. In the case of SPD where usually the component of the stresses of uniform compression is high, the formation and growth of cleavage cracks is partially or completely suppressed and, consequently, the conditions for fracture become more difficult. Using the schemes of the stress state with a higher component of the compressive stresses deformation of the solid is 'arrested', without fracture, and therefore the stage of developed deformation (fragmentation) and the SPD stage take place after the stage of macroscopic deformation. In the SPD stage, the active dissipative role is played by completely new processes

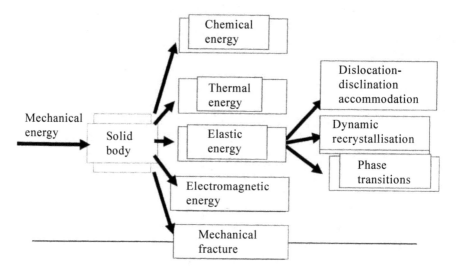

Fig. 5.5. Energy principle of the mechanical effect on the solid.

of dynamic recrystallisation, phase transformations, including amorphisation of the structure and/or generation of heat.

In [25] three possible scenarios ('the roadmaps') of the development of further events were proposed (Fig. 5.6). When the material is characterised by suitable conditions for the processes of dislocation and disclination rearrangement (for example in pure metals), the completion of the stage of developed deformation (fragmentation) is followed by low-temperature dynamic recrystallisation (upper branch in Fig. 5.6). The local regions of the structure are 'cleaned' to remove defects, and the new recrystallised grains are again characterised by the start of the process of plastic flow by dislocation and disclination modes. In this case, dynamic recrystallisation represents a powerful additional channel for the dissipation of elastic energy, and the stage of fragmentation and dynamic recrystallisation cyclically replace each other with increase of the strain in the fixed sections of the solid. When the mobility of the carriers of plastic deformation is relatively low (for example, in solid solutions or intermetallic compounds), a powerful additional channel of distribution of elastic energy is the phase transition (the lower branch in Fig. 5.6). In most cases, it is the 'crystal → amorphous state' transformation. Consequently, the plastic flow is localised in the amorphous matrix without the effects of strain hardening and build-up of high internal stresses. Evidently, there

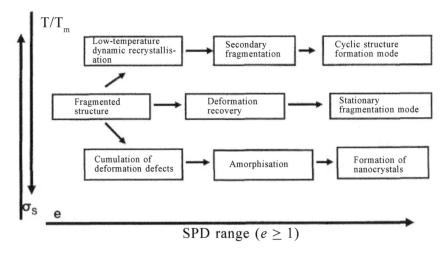

Fig. 5.6. The main scenarios (roadmap) of the development of structural processes in severe plastic deformation.

is an intermediate case (the central branch in Fig. 5.6) in which the additional dissipation channel may be the effective dislocation–disclination accommodation processes leading to the stabilisation of the fragmented structure with the development of SPD observed in certain experiments.

The transition from one 'roadmap' of structural rearrangement to another depends on the parameter T_{SPD}/T_m, where T_{SPD} is the temperature of SPD taking into account the possible effect of heat generation, and T_m is the melting point.

5.5. Low-temperature dynamic recrystallisation

Assuming the first scenario of the structural changes during SPD, one can conclude *a priori* at the same time that the processes of recrystallisation during SPD may take place even at the room temperature. Because the recrystallisation process (including dynamic recrystallisation) is mainly of the diffusion type [26], it should be concluded at the same time that the processes of diffusion and self-diffusion of the substitutional atoms, required for the formation of recrystallisation nuclei and their subsequent growth, may take place successfully in iron, nickel and other metals, and also in alloys based on these metals, at temperatures considerably lower than $0.3\ T_m$ at which the SPD experiments are usually carried out. At first sight, such a claim appears to incorrect, but the literature contains a large number of experimental confirmations according to which the dynamic recrystallisation during SPD actually takes place at the room and even lower temperatures [27, 28]. The dynamic recrystallisation process may take place both by the nucleation of new recrystallisation grains (discontinuous dynamic recrystallisation) and without the formation of new grain nuclei (continuous dynamic recrystallisation) [29].

A characteristic special feature of the structure of metallic materials after SPD is the existence of large local misorientations, corresponding to the high-angle grain boundaries [16]. It is generally believed that these boundaries determined to a large extent the unique properties of the materials after SPD. However, what is the physical nature of the formation of high-angle boundaries at high strains ? Unfortunately, it is not possible to provide a correct answer to this important problem using the model of developed plastic deformation, including the structural–kinetic approach and the concept of the mesodefects formed during deformation at relatively low true strains ($e \approx 1$–2) [22, 23]. The recent detailed review of this problem [30] also does not contain an answer to this problem.

In order to explain the resultant situation, SPD experiments were carried in [31] using a Bridgman chamber and a single-phase polycrystalline FeNi alloy with a crystal structure of the FCC solid solution.

Figure 5.7 shows the histograms of the size distribution of the grains and fragments after deformation for different numbers of total rotations N. The characteristic size of the structural element was the value $D = \sqrt{S}$, where S is the area of the image of the grain or fragment. Analysis of all resultant distribution shows the existence of two maxima (bimodal distribution) whose parameters change depending on the strain, but the bimodal nature of the distribution remains unchanged.

Using the energy concept of existence of the additional dissipation channels of the elastic energy in SPD [24], proposed by the authors of the present book, and also in a number of experiments (for example [27]] it was shown convincingly that the SPD processes at room temperature in the metals and solid solutions are accompanied by dynamic recrystallisation. In the electron microscopic analysis of the structure of the FeNi alloy after different deformation conditions in the Bridgman chamber the results showed distinctive features of the occurrence of similar processes (Fig. 5.8). The dark field electron microscopic images of the structure show clearly the individual

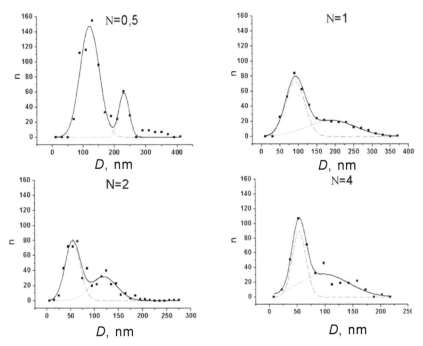

Fig. 5.7. Histograms of the size distribution of the grains and fragments and their division into two Gaussian distributions after severe plastic deformation with different values of the total rotations N for 50N alloy: — general histogram,--- – deformation fragments, ... – recrystallised grains.

grains having the shape of almost regular hexagons and the low density of the defects (grains 1 and 2 in Fig. 5.8). This indicates the formation of these grains in the process of dynamic recrystallisation by the migration of high-angle boundaries or co-operative rotation of the individual fragments in the conditions of the high concentration of the point defects and their gradients of the internal stresses [32]. At the same time, some grains (fragments) have an irregular shape and contain a high density of defects and extensive local distortion (grains 3 and 4 in Fig. 5.8) which evidently indicates that they formed as a result of the deformation fragmentation processes [22].

Each of the two types of grains, formed in the structure after the effect of SPD, should evidently be characterised by its own size distribution. In other words, the histograms, shown in Fig. 5.7, are in fact combined and consists of two histograms of the distribution of the grains and fragments characterised by different mechanisms of formation.

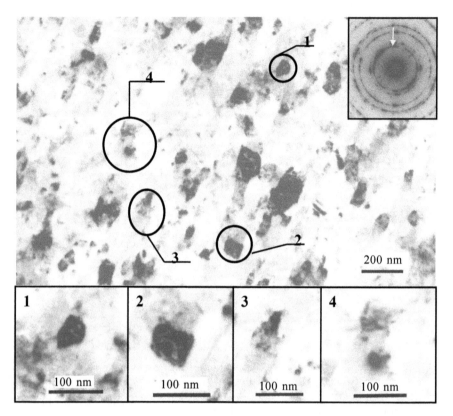

Fig. 5.8. Electron microscopic images of the structure and electron diffraction micrographs of 50N alloy after severe plastic deformation ($N = 2$); the dark field in separated reflection; 1, 2 – recrystallised grains; 3, 4 – dislocation fragments.

Figure 5.7 also shows the examples of dividing each experimentally determined bimodal histogram for different values of N into two Gaussian distributions, with the procedure described in detail in [33]. Analysis of the electron microscopic images shows that the distribution, indicated by the broken lines, corresponds to the fragments of deformation origin, and the distribution denoted by the dotted lines – to the recrystallised grains. Further, by calculating the relative areas below the Gaussian distribution experiments were carried out to determine the relative fraction of the volume, occupied by the deformation fragments C_{fr} and the recrystallised grains C_{rec} for each deformation mode (Fig. 5.9a) and also the mean size of the regions corresponding to the fragments D_{fr} and the recrystallised grains D_{rec} (Fig. 5.9b). It may be seen that with increase of N the value C_{fr} initially rapidly increases, reaching 0.8, and then smoothly

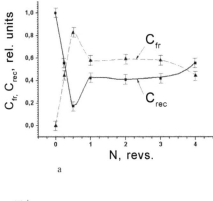

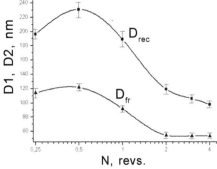

Fig. 5.9. Variation of the volume fraction C_{rec} and C_{fr} (a) and the average size D_{rec} and D_{fr} (b) for the recrystallised grains and deformation fragments in dependence on the number of total rotations N in the FeNi alloy.

decreases, down to the almost constant value of 0.6. The value C_{rec} which includes evidently not only the recrystallised but also initial grains, initially rapidly decreases to 0.2 and then increases to 0.4 which with increasing N remains almost completely constant. The authors of the book believe that the parameter $\kappa = C_{rec}/C_{tr}$ at $N > 1$ is an important structural constant and depends on the ratio of the material and the SPD parameters and determines to a large degree its physical–mechanical properties (in particular, the relationship between strength and plasticity). To confirm this, it is important to stress that the value of the microhardness measured in this work rapidly increases at $N < 1$ and remains constant at $N > 1$. SPD is accompanied by 'dynamic equilibrium between the structural transformations 'deformation fragments $\Leftrightarrow$ recrystallised grains. The dependences $D(N)$ in accordance with Fig. 5.9b have the same

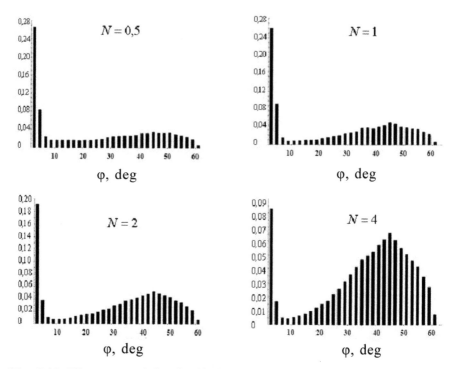

Fig. 5.10. Histograms of the distribution of the angle of misorientation φ in 50N alloy after different numbers of total rotations N, obtained by EBSD analysis.

form: the values of D_{fr} and D_{rec} smoothly decrease to 50 and 100 nm, respectively, and then at $N > 2$ remain almost completely constant.

Using EBSD analysis it was possible to determine the distribution of the angles of misorientation φ of the fragments and the grains, formed at different values of N (Fig. 5.10). With increasing N the characteristic feature is the large increase of the fraction of misorientation with φ > 20°, and at all values of N the nature of distribution of the misorientation remains almost completely constant: the maximum misorientation corresponds to the values of φ = 40–50°. This result contradicts the model of developed plastic deformation [22] in which the value of the maximum φ should increase monotonically with increasing N.

Figure 5.11 shows the results of measurements of the coercive force H_c (a) and microhardness HV (b) in dependence on the number of rotations N (the strain) in a 50N alloy. Both dependences H_c (N) and HV (N) show the same behaviour: there is a rapid parabolic increase at $N < 1$ and establishment of saturation at $N > 2$. The

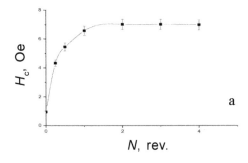

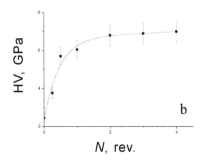

Fig. 5.11. Dependence of coercive force H_c (a) and microhardness HV (b) on the number of total rotations of the moving anvil N for the FeNi alloy.

similar form of the dependences has been detected many times for other solid solutions and pure metals not subjected to phase transformations during SPD [34], and is associated with the identical effect of the high internal stresses on the mobility of the walls of the ferromagnetic domains and the application of the external magnetic field and on the propagation of plastic shear in microindentation. Noting the identical form of the dependences in Fig. 5.11, it is however necessary to pay attention to the fact that the values of HV continue to increase slowly and monotonically also at $N > 2$.

Attention will be given to the correlation between the fraction of the dynamically recrystallised grains C_{rec} and the fraction of the high-angle grain boundaries α for the same structural state of the alloy in different stages of the SPD (for different values of N). The values of α were determined using the data shown in Fig. 5.10, with the restriction imposed on the minimum value of φ (α_{20} if $\varphi > 20°$; α_{30} if $\varphi > 30°$, and so on). The parametric dependences α (C_{rec}) were constructed for $N = \frac{1}{2} - 4$ and for $\alpha = \alpha_{20}-\alpha_{50}$. Figure 5.12 shows the experimental relationships for α_{40} (C_{rec}), and each point

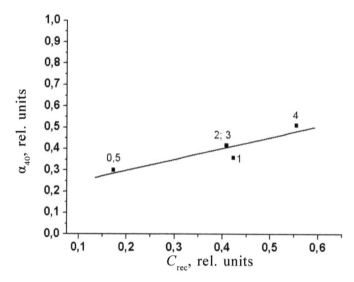

Fig. 5.12. Linear correlation of the parametric dependences α_{40} (C_{rec}) for the FeNi alloy; C_{rec} is the volume fraction of recrystallised grains, α_{40} is the number of the grain boundaries with the misorientation angle $\varphi > 40°$; the numbers indicate the number of the total rotations N; correlation coefficient $r = 0.912$.

corresponds to a specific value of strain N. The linear correlation coefficient r is determined by the standard procedure [35] and in this case is equal to 0.912. In other words, with the increase of the magnitude of SPD there is a strict linear correlation between the increase of the number of the grains in the structure formed during dynamic recrystallisation and the increase of the fraction of the high-angle grain boundaries in the structure ($\varphi > 40°$). This also unambiguously confirms that, the authors believe, the process of dynamic recrystallisation (and not deformation fragmentation) is responsible for the formation of a relatively large number of the high-angle grain boundaries in the structure of the material subjected to SPD. Two important consequences should be noted.

1. The maximum of the distribution of the high-angle grain boundaries is obtained for the misorientation angles of $\varphi = 40$–$50°$ which correspond to the highest mobility of the high-angle boundaries and, consequently, the most efficient growth of recrystallisation nuclei [26].

2. A decrease of the minimum value of φ in the construction of the parametric dependences $\alpha(C_{rec})$ ($\alpha_{50} \to \alpha_{20}$) is accompanied by the decrease of the linear correlation coefficient which indicates undoubtedly the increasing contribution of the processes of dynamic

fragmentation to the formation of the grain boundaries with the average and small misorientation angles.

Comparing the nature of the variation of the volume fraction of the recrystallised grains C_{rec} and the deformation fragments C_{fr} (Fig. 5.9a) with the variation of the physical and mechanical properties (Fig. 5.11) with increasing magnitude of the SPD, determined by the parameter N, it may be concluded that there is a specific relationship between them. In fact, the rapid increase of the volume fraction of the deformation fragments characterised by the high values of the internal stress fields, corresponds to the same rapid increase of the values of H_c and HV at $N < 2$. The 'dynamic equilibrium' observed subsequently at $N > 1$ between the regions of the deformed crystal, occupied by the deformation fragments of the recrystallised grains and corresponding to the stabilisation of the value $C_{fr} \approx 0.5–0.6$, leads (with a certain 'delay' with respect to the parameter N) to the stabilisation of the values H_c and HV. Thus, it may be assumed that the structurally sensitive strength and magnetic properties of the single-phase materials, subjected to SPD, are determined by the ratio of the volume fractions for the deformation fragments and the recrystallised grains $k = C_{fr}/C_{rec}$. As the value of k increases the strength (microhardness) and magnetic coercivity of the ferromagnetics increase. In this work and in investigations carried out using commercial purity iron [33] the value of k was close to unity. It may also be assumed that the increase of the volume fraction of the recrystallised grains (lower values of k) may correspond to the increase of the ductility of the materials subjected to the SPD.

Thus, the structure after SPD can be regarded as a 'two-phase mixture' consisting of fragments and recrystallised grains. In addition, the presence in the structure of the materials subjected to the SPD of a high density high-angle boundaries of the grains is determined mostly by the occurrence of the processes of dynamic recrystallisation and not of deformation fragmentation, as assumed previously. The strength (microhardness) and magnetic coercivity of the ferromagnetic single-phase alloys after SPD increases with increase of the value of $k = C_{fr}/C_{rec}$, where C_{fr} and C_{rec} is the volume fraction of the deformation fragments and the recrystallised grains in the structure of the deformed material.

5.6. Amorphisation and crystallisation during SPD

As mentioned previously, the additional channel of dissipation in the intermetallic compounds in other materials with a low mobility of the dislocations may be the amorphisation process. The most typical example is the titanium nickelide in which the transition to the amorphous state was observed after SPD in the Bridgman chamber [36] and after cold rolling [37]. Evidently, the crystal containing a high concentration of linear and point defects is thermodynamically unstable with respect to the transition to the first stage, especially if the difference between the free energies of the crystalline and amorphous state is not large.

Naturally, it is necessary to answer the question what would take place if SPD is carried out on the initially amorphous state produced, for example, by melt quenching or some other method ? Taking into account these considerations, the amorphous state should remain amorphous. However, as explained in [38–40], SPD is accompanied by nanocrystallisation: there are nanocrystals with the size of approximately 10–20 nm distributed homogeneously or heterogeneously in the amorphous matrix.

Amorphous alloys of the metal–non-metal type

The formation at room temperature of the nanocrystals with the size of up to 20 nm, uniformly distributed in the entire volume of the amorphous matrix, is difficult to explain on the basis of the classic consideration of the thermally activated nature of the crystallisation processes. In [41] the authors attempted to analyse in detail the special features of the structure and properties under the effect of SPD on a number of amorphous alloys of the metal–metalloid type. Basically, investigations were carried out on the $Ni_{44}Fe_{29}Co_{15}Si_2B_{10}$ alloy produced by melt quenching. However, in addition in [41] investigations were carried out on $Fe_{74}Si_{13}B_9Nb_3Cu$ (Finemet), $Fe_{57.5}Ni_{25}B_{17.5}$, $Fe_{49.5}Ni_{33}B_{17.5}$ and $Fe_{70}Cr_{15}B_{15}$ alloys. Figure 5.13 shows the X-ray diffraction diagrams of the Ni–Fe–Co–Si–B amorphous alloy in the initial state (a), after $N = 4$ at 293 K and 77 K (b), and after $N = 8$ at the same deformation temperature (c) (N is the number of total rotations in the Bridgman chamber). It may be seen that after SPD crystallisation processes start to take place in the alloy and these processes are far more distinctive after deformation at room temperature. A special computer program was used to determine the volume fraction and the size of the crystalline phase in the case of X-ray diffraction diagrams shown in Fig. 5.13. For example, for $N = 4$ and $T = 293$ K the fraction of the crystalline phase α and the average size of the crystals d were equal to 8% and 3 nm, respectively. It is

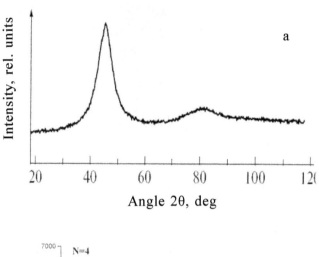

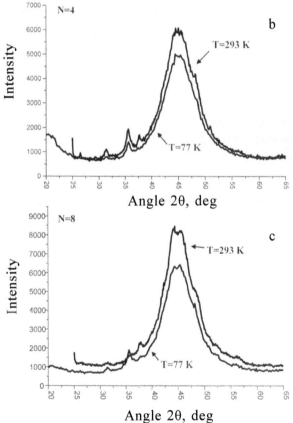

Fig. 5.13. X-ray diffraction diagrams of the $Ni_{44}Fe_{29}Co_{15}Si_2B_{10}$ amorphous alloy in the initial condition after melt quenching (a), after $N = 4$ at 293 K and 77 K (b) and after $N = 8$ at the same deformation temperatures (c).

interesting to note that the values of these parameters correspond almost completely to those obtained after $N = 0$, but at $T = 77$ K ($\alpha = 6\%$ and $r = 2$ nm). Electron microscopic images confirm these results quantitatively and qualitatively.

Figure 5.14 shows the variation of the microhardness HV of the amorphous and partially crystallised alloys (preliminary annealing of the amorphous state at the temperature higher than T_{cr}) in dependence on the strain N in the Bridgman chamber at different temperatures. Attention should be given to the large difference in the form of the curves in the initially amorphous and initially partially crystalline states. In the former case, HV decreases rapidly in the initial stages of the SPD ($N = 0.5$) and this is followed by a monotonic increase, and both the effect of the decrease and the effect of subsequent growth are more distinctive at $T = 293$ ($- 2.8$ and $+2.1$ GPa at 293 K and -0.80 GPa and $+1.1$ GPa at 77 K). As regards the initial partially crystalline alloy, its HV value was 0.75 GPa higher than in the initial amorphous alloy, but after SPD there was initially a large and subsequently small decrease of the values of HV without any extreme changes to the value almost completely identical with the HV value for the amorphous non-deformed state.

The region of the rapid decrease of HV is characterised by the inhomogeneous plastic deformation with the formation of thick shear bands which is typical of all amorphous alloys at temperatures

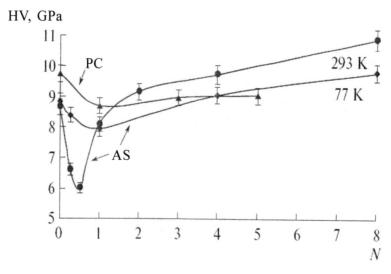

Fig. 5.14. Measurement of the microhardness HV of the amorphous (AS) and partially crystallised (PC) $Ni_{44}Fe_{29}Co_{15}Si_2B_{10}$ alloy in dependence on the value of N after severe plastic deformation at 293 and 77 K.

considerably below the point of transition to the crystalline state [42]. The local shear bands were detected by transmission electron microscopy on the deformed specimens as a result of the fact that they contained crystallisation effects (Fig. 5.15). Otherwise, the contrast on the electron microscopic images may be only of the absorption nature and it is necessary to prepare the specimens for electron microscopic studies and then deform the specimen. The theoretical estimates show that the local increase of temperature in the shear bands may reach 500°C [43]. In this case, the local temperature in the plastic shear zone may exceed the crystallisation temperature of the amorphous alloy (in the present case $T_{cr} = 410$°C) and result in the formation of the primary FCC phase in the shear bands. This is confirmed by the results obtained in [44] where it is shown that as the crystallisation temperature of the amorphous alloy increases, the volume fraction of the crystalline phase, formed during SPD at room temperature (Fig. 5.16) decreases.

In later stages of SPD ($N > 1.0$) the deformation pattern dramatically changes. The shear bands are no longer detected. Instead, the structure shows the nanoparticles of the crystalline phase with the size of up to 10 nm homogeneously distributed throughout the entire volume of the specimen (Fig. 5.17). It may be concluded that the process of plastic deformation of the amorphous alloys ceased to be strongly localised, inhomogeneous, and in all likelihood, transformed to the 'quasi-homogeneous' process. This nature of the plastic flow is typical of the amorphous alloys at very

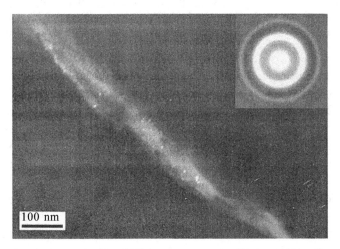

Fig. 5.15. Electron microscopic image of the crystallised local shear band in the $Ni_{44}Fe_{29}Co_{15}Si_2B_{10}$ amorphous alloy after severe plastic deformation ($N = 0.5$; $T = 293$ K).

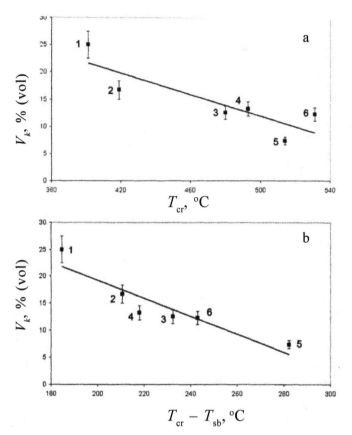

Fig. 5.16. Correlation between the values of V_k and T_{cr} for the amorphous alloys Fe–Ni–B (1), Fe–Ni–B (2), Fe–Cr–B (3), Fe–Co–Cr–B (4), Fe–B–Si (5) and Fe–Cr–Zr–B–C (6) without taking into account (a) and taking into account (b) different dissipative capacity of the shear bands; correlation coefficient $r = 0.74$ (a) and $r = 0.86$ (b).

high temperatures, close to the glass transition temperature in the conditions of the large decrease of the dynamic viscosity of metallic glass [42]. In this case, it is possible to reach the similar 'softened' state at room temperatures which is not possible at 77 K. Evidently, in this case, we are concerned with the occurrence of the completely new structural mechanism of plastic deformation of the amorphous alloys which is observed only in the SPD conditions.

One of the possible explanations of this development of the events is as follows [45]. During propagation of a shear band in the amorphous matrix in the course of SPD the temperature of the matrix constantly increases and the temperature of the front of the matrix is always maximum (Fig. 5.18). This is followed by the

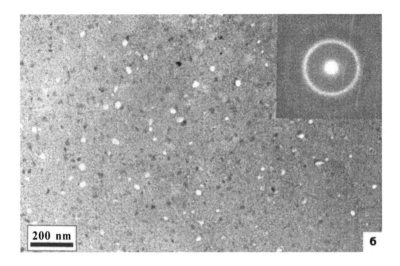

Fig. 5.17. Dark field image of the $Ni_{44}Fe_{29}Co_{15}Si_2B_{10}$ amorphous alloy after $N = 8$ at 77 K.

start of the phase of propagation of the band in which the local temperature of the front reaches the crystallisation temperature (II in Fig. 5.18). A nanocrystal forms at the front of the growing band and efficiently inhibits the plastic flow zone because the resultant crystal has the nanosize and is not capable of dislocation plastic flow. There are two possible variants of the further development of the process. Firstly, under the effect of the shear band the nanocrystal will accumulate a high level of elastic stresses so that the amorphous matrix will be characterised by the nucleation of a new shear band by the elastic accommodation mechanism (III in Fig. 5.18). In this case, the plastic flow process will take place by the relay mechanism, generating in the shear band nanocrystals equidistantly distributed along the trajectory of movement of shear bands in the amorphous matrix. This is confirmed by the electron microscopic image in Fig. 5.19 which in fact shows the changes of the equidistantly distributed nanocrystals formed as a result of the SPD. Secondly, the branching of the shear bands, inhibited by the frontal formation of the nanocrystals, may take place. This process which resembles to some extent the multiplication of the dislocations at non-intersected particles is shown schematically in Fig. 5.20. As a result of this 'self-inhibition'of the shear bands at the frontal nanocrystals the de-localisation of the inhomogeneous plastic flow takes place in the later stages of the SPD. In fact, the observed effect of transition to

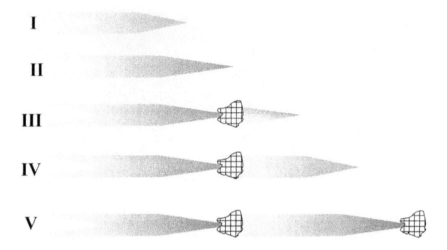

I

II

III

IV

V

Fig. 5.18. The mechanism of 'self-blocking' of the shear band, propagating in the amorphous matrix.

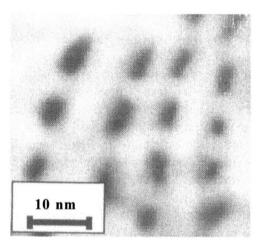

10 nm

Fig. 5.19. The chains of equidistant nanocrystals, formed in severe plastic deformation of the Fe–Ni–B amorphous alloy. Transmission electron microscopy.

homogeneous nanocrystallisation at the shear bands indicates that the plastic flow is characterised by the high volume density of the shear bands and, consequently, the homogeneous formation of nanocrystals in 'thinner' shear bands.

Thus, nanocrystallisation is a consequence of the local generation of heat as a result of the plastic deformation processes. The thermally activated nature of nanocrystallisation and, consequently,

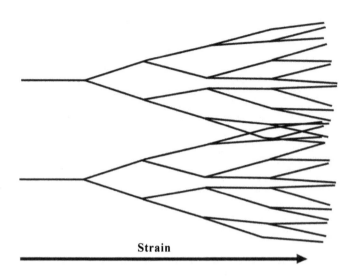

Strain

Fig. 5.20. The mechanism of multiplication of the shear bands, interacting with the frontal nanocrystals.

its formation as a result of the local increase of temperature is also indicated by the fact that the structural state and, correspondingly, the microhardness after $N = 4$ at room temperature correspond accurately to the structural state and the microhardness after $N = 8$ in deformation at 77 K [41]. In other words, the higher strength values compensate the temperature deficit in the diffusion nanocrystallisation processes.

Examination of the SPD of the partially crystallised alloy resulted in unexpected data [46]. A series of X-ray diffraction patterns, shown in Fig. 5.21, shows clearly that the partially crystalline state, formed after annealing of the amorphous alloy, again becomes amorphous when the strain is increased to $N = 5$. The electron microscopic experiments unambiguously confirmed this tendency: there is a rapid decrease of the size of the nanocrystals without affecting their bulk density. The unusual evolution of the amorphous–nanocrystalline structure with increasing N is clearly confirmed by the dependence of the mean size of the nanoparticles and their bulk density with the increase of N in the partially crystallised alloy (Fig. 5.22). At an almost constant number of the nanoparticles in the unit volume, the crystalline phase disappears as a result of a large decrease of the size of the nanoparticles. In other words, this takes place as a result of their 'dissolution' in the amorphous matrix. The 'dissolution' of the nanoparticles during SPD can be described even more accurately

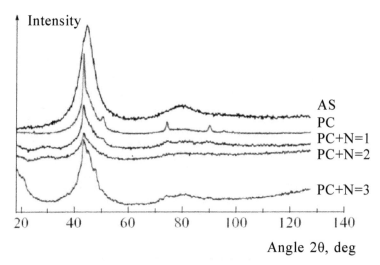

Fig. 5.21. Evolution of the X-ray diffraction diagrams of the partially crystallised (PC) state of the $Ni_{44}Fe_{29}Co_{15}Si_2B_{10}$ alloy after severe plastic deformation with different values of N ($T = 293$ K).

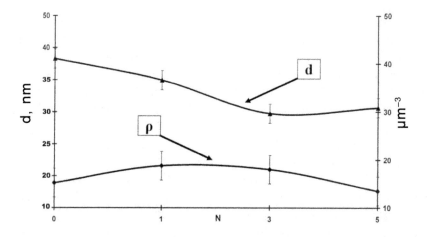

Fig. 5.22. Dependence of the average size (d) and bulk density (ρ) of the nanoparticles of the crystalline phase in the initial partially crystallised alloy $Ni_{44}Fe_{29}Co_{15}Si_2B_{10}$ on the value of N in severe plastic deformation ($T = 290$ 3K).

by comparing the histograms of the distribution of the particles of the crystalline phase obtained after different conditions of the SPD (Fig. 5.23). It may be seen that with increasing N the 'tail' in each consecutive histogram disappears. The 'tail' corresponds to the largest nanoparticles (crosshatched in the histograms shown in Fig. 5.23).

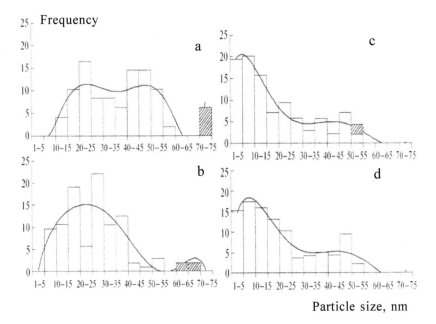

Fig. 5.23. The size distribution histograms of the nanocrystals observed in different stages of severe plastic deformation of the partially crystallised $Ni_{44}Fe_{29}Co_{15}Si_2B_{10}$ alloy; $N = 0$ (a), $= 1$ (b), $= 3$ (c), $= 5$ (d), the dimensional fractions which disappear during increasing strain are indicated by crosshatching.

Thus, the clearly contradicting (at first sight) results have been obtained. On the one hand, the SPD of the $Ni_{44}Fe_{29}Co_{15}Si_2B_{10}$ amorphous alloy results in partial transition of the alloy to the crystalline (more accurately, nanocrystalline) state. On the other hand, the SPD of the same partially crystallised alloy results in the dissolution of the crystalline phase, i.e., there is a tendency to return to the initial amorphous state. The observed contradiction is apparent and can be logically explained taking into account the specific features of the structural processes taking place during the SPD.

The general energy concept of the SPD will now be discussed [47]. A large amount of elastic energy is added to the solid during deformation. In this case, the possible dissipation channels should include plastic deformation, phase transformation and the generation of heat. Crystallisation may be caused by both the local increase of temperature and the presence of high local stresses in the amorphous matrix. The stresses stimulate the processes which depend on temperature, and as the stresses increase the temperature at which the thermally activated crystallisation process takes place decreases. In addition, it should also be taken into account that the activation energy of the crystallisation process Q^* is lower than that of the conventional process because of the considerably higher concentration of the regions of the excess free volume in the shear bands [48]. Finally, it is also important to take into account another important feature: the local atomic structure of the amorphous matrix in the shear band may differ from the 'classic' structure for the amorphous state. It is fully possible that the amorphous matrix contains (even prior to crystallisation) the deformation-induced areas with higher correlation in the distribution of the atoms – the nuclei of the crystalline phase with a considerably different degree of the composite and topological short-range order. This is indirectly indicated by the experimental results obtained in [49] in which it is shown that the chemical composition of the crystalline phase in the amorphous alloy based on aluminium greatly differs after conventional annealing and after SPD.

Thus

$$Q^* = Q_k - G\tau - \Delta Q_{fv} - Q_{so} \qquad (5.3)$$

where Q^* is the effective activation energy of crystallisation in the shear band, Q_k is the activation energy of crystallisation as a result of thermal fluctuations, G is the shear modulus, τ is the shear stress in the region of the plastic shear zone, ΔQ_{so} is the contribution to the decrease of the activation energy of crystallisation associated with the presence of the short-range order (topological and/or compositional) in the zone of the shear band, and ΔQ_{fv} is the contribution to the decrease of the activation energy of crystallisation as a result of a significant enrichment of the shear bands by the excess free volume.

It should be mentioned that the formation of crystals in the shear bands takes place in the process of SPD, not after completion of SPD. Consequently, the shear bands newly formed in the amorphous matrix start to interact with the crystals formed in any stages of the SPD. A similar interaction may take place by several mechanisms [50]: inhibition of the shear bands at the particles of the crystalline phase, intersection or bending of such particles by the shear bands, and also the primary and secondary accommodation effects. In any case, a similar interaction may cause the formation of dislocations in the crystalline particle. The highest dislocation density will be evidently observed in the vicinity of the interphase boundary where the effect of the shear bands is the strongest. Finally, a moment arises in which the dislocation density in the boundary zone will be extremely high and the zone (or the entire crystalline particles) transfer spontaneously to the amorphous state because the free energy of the area of the highly defective crystal is higher than the free energy of the amorphous state. This will be perceived as the 'dissolution' of the crystals under the effect of the shear bands actively operating in the amorphous matrix during SPD. In particular, this 'dissolution' process was detected in the SPD of a partially crystallised amorphous alloy (Figs. 5.22 and 5.23). It should be remembered that the deformation 'dissolution' of the crystals can hardly continue to the end. It is well-known [51] that the very small crystalline particles (smaller than 10 nm) are not capable of storing dislocation type defects because of the presence of very high imaging forces. In this case, this means that the nanocrystalline particles smaller than 10 nm, located in the amorphous matrix, will not 'dissolve' during SPD. In other words, the nanocrystals smaller than 10–20 nm, formed in the amorphous matrix, will be structurally stable and will remain unchanged throughout the long SPD stages. These results can be used to explain the fact that in all investigations without exception where the process of precipitation of the crystals during the SPD was detected, the size of these crystals was always smaller than 20 nm.

At the same time, as shown by the experiments, there may be cases in which the balance between the formation of the crystals on the shear bands and subsequent deformation 'dissolution' may be disrupted. In this case, SPD is accompanied by the operation of the cyclicity process, and the structural states with a large or small volume fraction of the nanocrystals in the amorphous matrix periodically replace each other with increasing strain [52].

Ti–Ni–Cu amorphous alloys

Recently, special attention has been paid to alloys based on titanium nickelide characterised by the shape memory effect [53]. The results show that the TiNi intermetallic compound in the process of SPD in shear under pressure in the Bridgman chamber or during cold rolling can be transferred partially or completely to the amorphous state [54]. Subsequently, this effect, typical of the titanium nickelide, has been confirmed many times by other investigators [55, 56]. Figure 5.24 shows the linear dependence of the volume fraction of the amorphous phase, formed in different alloys in the vicinity of the TiNi composition on the strain in cold rolling [57].

At the same time, studies have been published in which an alloy based on titanium nickelide $Ti_{50}Ni_{25}Cu_{25}$ was produced in the amorphous state by melt quenching and then subjected to SPD in the Bridgman chamber. At a certain stage of deformation there was a transition from the amorphous to nanocrystalline structure [59]. Thus, on the one hand, SPD results in the occurrence of the crystal→amorphous state phase transition, and on the other hand the amorphous state→crystal (nanocrystal) phase transition. This evident contradiction, which was previously discussed for the amorphous alloys of the metal–non-metal type, was used by the authors of [60] as an impetus to carry out a detailed and systematic investigation of the structural and phase transformations in alloys based on the titanium nickelide under the effect of SPD in the Bridgman chamber with the variation of the chemical composition, initial structure, and also the temperature and magnitude of the SPD. Experiments were carried out on the $Ti_{50}N_{25}Cu_{25}$ alloy which could be in both the crystalline and amorphous state prior to deformation in the Bridgman chamber. The amorphous state can be produced by vacuum quenching from the melt by spinning at a speed of 10^6–10^7 deg/s.

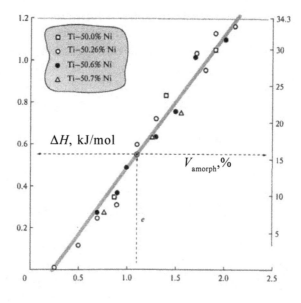

Fig. 5.24. Variation of the amount of the amorphous phase with increasing strain in cold rolling of Ti–Ni alloys.

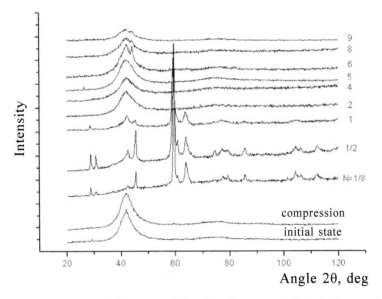

Fig. 5.25. Profiles of the X-ray diffraction diagrams of the initial amorphous alloy $Ti_{50}Ni_{25}Cu_{25}$ (in), after hydrostatic compression ($P = 4$) without shear and after shear under pressure with different numbers of rotations ($N = 1/8$, 0.5, 1, 2, 4, 6, 8).

The results obtained in [60] confirmed unambiguously that phase transformations of different type take place in the $Ti_{50}Ni_{25}Cu_{25}$ alloy, produced by melt quenching, as a result of SPD (Fig. 5.25).

Applying only the hydrostatic pressure without shear deformation already results in the formation of a small amount of the B19 crystalline phase in the amorphous matrix. In the initial stage of SPD ($N = 1/8$) the volume fraction of the crystalline phase (type B19 and B2) rapidly increases (to ~70%). The extremely high value of the volume fraction of the crystals (~80%) was observed after $N = 0.5$ and then (after $N = 1$) the amount of the crystalline phase rapidly decreased (~30%) (Fig. 5.26). This tendency of the decrease of the amount of the crystalline phase remained unchanged with increasing strain, and after $N = 2$ the structure contained almost no crystalline phase. This structure is indicated by the X-ray diffraction diagram shown in Fig. 5.27. It is typical of the amorphous state of the solid, and only the unusual appearance of some electron microdiffraction patterns and dark field high-resolution electron micrographs indicate the presence of a small amount of the nanocrystalline phase. Deformation at $N = 4$ resulted in the complete 'dissolution' of the nanocrystals and the formation of the amorphous state which only

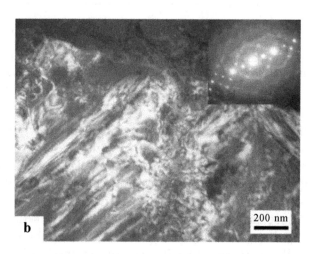

Fig. 5.26. Electron microdiffraction photographs and dark field electron microscopic images of the structure of the $Ti_{50}Ni_{25}Cu_{25}$ alloy after $N = 1$ in the (110)B2 reflection; the axis of the zone [001]B2 (a) and the reflection (100)B19; the axis of the zone [010]B19 (b).

in some details differs from the initial amorphous state, produced by melt quenching. A further increase of the SPD value ($N = 5$) leads again to the appearance of a small amount of the crystalline phase in the structure detected only on the electron microscopic images. However, even a large increase of the strain ($N = 6$) 'fixes' the crystalline phase both in X-ray diffraction examination and electron microscopy (Fig. 5.28). Theoretical analysis of the X-ray diffraction diagram with the split halo together with the electron

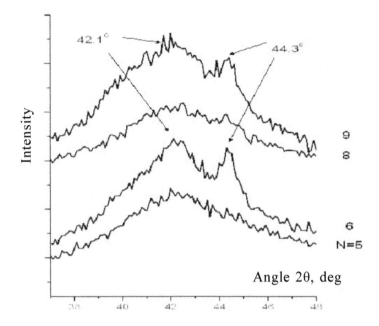

Fig. 5.27. Evolution of the main (first) halo of the X-ray diffraction diagrams of $Ti_{50}Ni_{25}Cu_{25}$ alloy at different magnitudes of severe plastic deformation. The number of rotations in the Bridgman chamber is indicated in the graph.

microscopic data indicate that the two-phase amorphous-crystalline structure forms in this case. A further increase of the intensity of SPD results in repeated 'dissolution' of the crystalline phase ($N = 8$) and in its subsequent appearance ($N = 9$). These results are clearly demonstrated by the graph in Fig. 5.29 which shows the dependence of the volume fraction of the crystalline phase in the initial amorphous alloy $Ti_{50}Ni_{25}Cu_{25}$ on the number of revolutions N in the Bridgman chamber. The results, shown in the graph, were obtained by the X-ray diffraction analysis method (a large volume fraction of the crystalline phase) and by transmission electron microscopy (a small volume fraction).

Evidently, in the investigations carried out in [52] it was possible to detect for the first time three cycles of the mutual amorphous – crystalline phase transitions. Previously, a similar effect was observed in the experiments with the mechanical activation of a powder of Co_3Ti intermetallic compound [61]. The similar cyclic process, i.e., the tendency to the phase transition of the initial amorphous phase to the crystalline state and vice versa in further stages, the existence of the completely opposite tendency in specific stages of the SPD

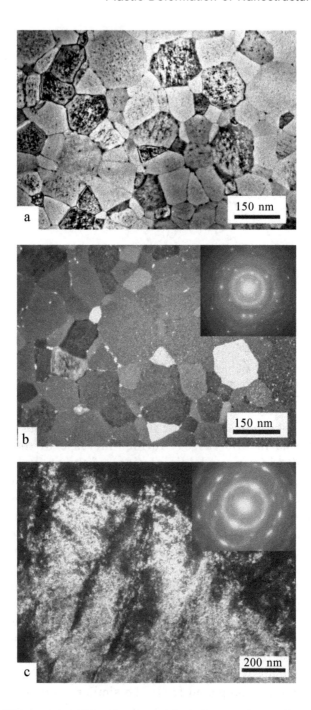

Fig. 5.28. Electron microdiffraction patterns and electron micrographs of the structure of the $Ti_{50}Ni_{25}Cu_{25}$ alloy, corresponding to severe plastic deformation at $N = 6$; a, c) the bright field; b) the dark field.

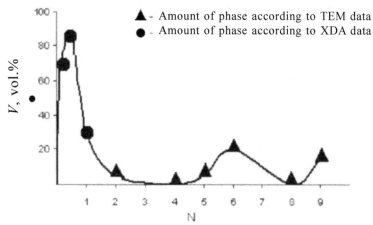

Fig. 5.29. Dependence of the volume fraction of the crystalline phase on the magnitude of deformation of the $Ti_{50}Ni_{25}Cu$ amorphous alloy.

explains without contradiction the apparent contradiction in the experimental results of the studies [54–56], on the one hand, and the studies [58–60] on the other hand.

It will be attempted to explain what is the principle of this cyclic process. The 'initial amorphous state → crystal (nanocrystal)' phase transition is evidently associated with two reasons. Firstly, as mentioned previously, the applied hydrostatic pressure is capable of stimulating the phase transition in which the equilibrium phases characterised by a lower specific volume. The fact that the $Ti_{50}Ni_{25}Cu_{25}$ amorphous alloy transforms partially to the crystalline state without shear deformation, and only as a result of the application of hydrostatic pressure (P = 4 GPa), confirms this hypothesis. Secondly, the appearance of additional channels of dissipation of elastic energy, characteristic of the SPD processes [22], results in the conditions of the strongly localised plastic yielding in the shear bands in the local release of thermal energy and in the corresponding local increase of the temperature of the amorphous matrix. In these conditions, the shear bands, as already mentioned, are capable of crystallising.

After the formation of the crystalline phase in SPD, the phase is subjected to very severe plastic deformation. One of the acting energy dissipation channels in the conditions of low mobility of the dislocations may be the amorphisation [22]. A similar phase transition (solid phase melting) becomes possible only when the free energy of the crystal, containing the colossal density of the defects (vacancies, dislocations, disclinations, fragment boundaries, etc.) becomes higher than the free energy of the disordered state of the system. The negative volume effect becomes here of secondary importance and is not controlling. These transitions were observed in experiments mostly in the intermetallic compounds, in particular, in the titanium nickelide [53–57].

Further with the development of the SPD the pattern is evidently repeated, but with certain special features associated with the increasing elastic energy and high (gigantic) internal stresses. Evidently, the described tendency to the cyclic transformations is common for the behaviour of metallic materials susceptible to amorphisation, in the course of high-energy effects, in particular, in SPD, carried out by different methods.

An important special feature of the processes taking place during SPD is the inhomogeneity of these processes, typical of any mode of plastic deformation [9]. For this reason, the cyclic amorphous–crystalline phase transitions are 'mismatched' in the volume of the deformed material and take place in different microvolumes at different deformation parameters. This may be used to explain the result that the states with the limiting content of either the amorphous or crystalline phase are recorded only in a very small number of cases. In any stage of severe plastic deformation there are local areas of the matrix which either 'outstrip' the adjacent regions or 'lag' behind them. As the strength increases, this mismatch effect becomes greater. Finally, we can reach a specific dynamic equilibrium in which it will not be possible to record any changes in the phase composition within the framework of the given averaging scale.

It is very interesting to analyse special features of the deformation amorphisation and crystallisation of the same material with the variation of its initial state which will make it possible to construct a single structural model of cyclic phase transformations during SPD.

Investigations were carried out into the structural and phase transformations for the variation of the magnitude of SPD using the Bridgman chamber in $Ti_{50}Ni_{25}Cu_{25}$ alloy which in contrast to the investigations carried out in [62] was in the crystalline state (not amorphous) prior to the start of the deformation experiments. The main results of these investigations can be summarised as follows:

– the investigated $Ti_{50}Ni_{25}Cu_{25}$ alloy has the initial crystalline structure represented mainly by the plate-shaped martensite B19. During SPD the plates turn round, disintegrate, become smaller and finally, completely disappear (Fig. 5.30);

– the amorphisation of the alloy starts already after $N = 0.25$ (Fig. 5.31). After deformation $N = 1$ ($e = 2.15$) there is the distinctive mass degradation of the plate-shaped structure and transition to the amorphous state;

– the degradation of the martensite plates is accompanied by the plastic deformation of the resultant amorphous phase as a result of which, starting at $N = 0.5$ ($e = 1.80$), electron microscopic images show the appearance of the nanocrystalline B2-phase with the size of the individual particles up to 10 nm (Fig. 5.32). In addition, there are sometimes globular regions of the B2-phase with the size of approximately 300 nm containing a large number of defects of the deformation origin;

– further deformation $N = 2$–4 ($e = 2.51$–2.90) is structurally characterised by the superposition of the amorphous phase and the nanocrystals of the B2-phase which often form in the shear bands of the amorphous matrix (Fig. 5.33);

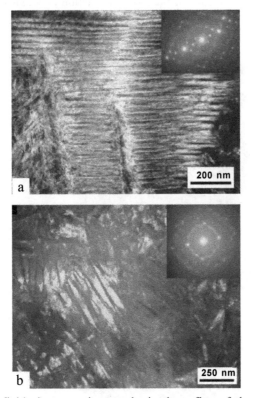

Fig. 5.30. Dark field electron micrographs in the reflex of the B19 martensitic phase and the appropriate electron diffraction patterns prior to SPD (a) and after SPD ($N = 1$) (b).

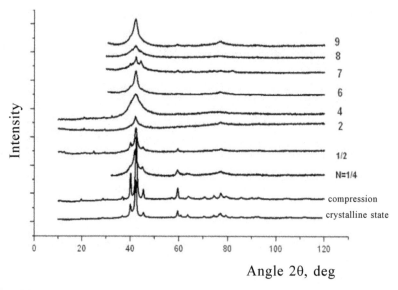

Angle 2θ, deg

Fig. 5.31. Complete profiles of the X-ray diffraction diagrams of the $Ti_{50}Ni_{25}Cu_{25}$ alloy in different stages of the experiments

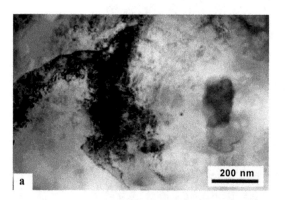

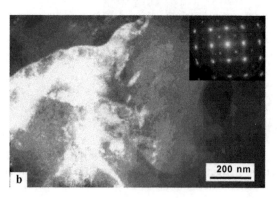

Fig. 5.32. Bright field (a) and dark field (in the reflection of the B2-phase) (b) electron micrographs indicating the presence of the B2-phase after SPD ($N = 0.5$).

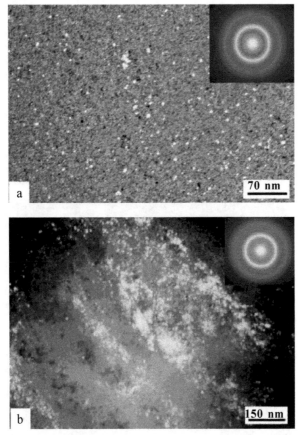

Fig. 5.33. Electron micrographs (dark field) of the nanocrystals of the B2-phase distributed uniformly in the volume (a) and in the shear bands (b) after SPD ($N = 2$).

– at later stages of deformation after $N = 6$ ($e = 3.5$) electron microscopic studies show the state of the local instability of the B2-phase which is an intermediate stage of the martensitic transformation B2 → B19;

– the X-ray diffraction diagram of the later stages of deformation after $N = 7$ ($e = 4.0$) are already characterised by the presence of wide maxima of the B19-phase at the background of the X-ray amorphous state;

– the X-ray diffraction diagrams in the final stages of deformation ($N = 9$, $e = 5.3$) are again completely amorphous, but the electron microscopic images indicate the presence of the nanocrystals of the B2-phase (Figs. 5.31 and 5.34).

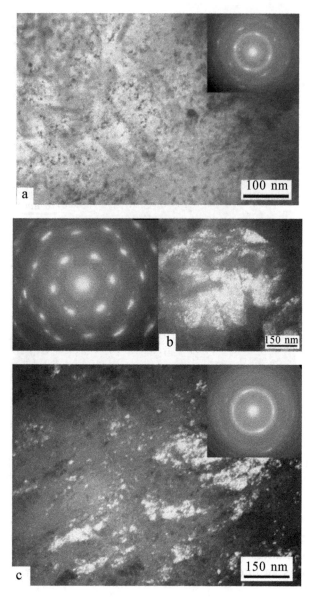

Fig. 5.34. Electron micrographs (bright field (a) and dark field (b, c) in reflections of the crystalline phases) in the appropriate microdiffraction patterns in different stages of the SPD: $N = 4$ (a), 6 (b), 9 (c).

Thus, as in the case of the initially amorphous $Ti_{50}Ni_{25}Cu_{25}$ alloy [53], the $Ti_{50}Ni_{25}Cu_{25}$ crystalline alloy is subjected to increasing SPD in the Bridgman chamber is characterised by the following cyclic sequence of the phase transitions:

$$B19 \rightarrow AS \rightarrow B2 \rightarrow B19 \rightarrow AS \rightarrow B2,$$

where the AS is the amorphous state of the solid state.

In the literature, it is assumed that the periodicity of the structural changes in SPD in a general case is determined by the activation of different dissipation (relaxation) channels of elastic energy stored by the material during deformation [24, 25]. Evidently, the special features of the change of the structure in the $Ti_{50}Ni_{25}Cu_{25}$ crystalline alloy observed during SPD are associated with the special features of the occurrence of direct and reversed phase transformations of both the diffusion and martensitic type. Figure 5.35 shows the scheme providing information on the nature of the cyclic transitions in SPD from the crystalline to amorphous state and then from the amorphous to nanocrystalline state with the subsequent periodic repetition of the processes but already at the nanoscale level.

The authors believe that the most interesting aspect is the explanation of the structural mechanism of the phase transition during SPD from the crystalline to amorphous state. In [62] experiments revealed the rotation, distortion and disintegration of the initially regularly distributed plates of martensite of the B19-phase in the process of SPD. However, the final stage of amorphisation – 'dissolution' of nanosized 'fragments' of martensite plates remains speculative. Evidently, one of the methods of verifying this assumption is the experimental observation of the dissolution of the nanosized crystals during deformation or also computer simulation

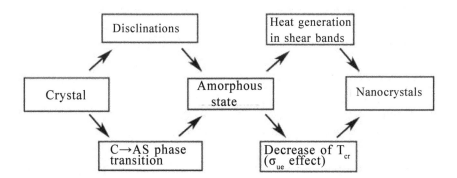

Fig. 5.35. Diagram of the processes leading to the transition from the crystalline to amorphous state and further the amorphous state of the crystalline state during SPD: C – crystal, AS – amorphous state, T_{cr} – the temperature of transition of the amorphous to crystalline state, σ_{ue} – the stress of the uniform (hydrostatic) compression.

of the process of solid-phase dissolution in the course of shear deformation in the uniform compression conditions.

Thus, it can be concluded that both in the deformation of intermetallic compounds and complex phases, susceptible to amorphisation during SPD, and in the deformation of the initially amorphous alloys SPD is accompanied by consecutive transitions from the amorphous or crystalline state and, vice versa, from the crystalline to amorphous state. Consequently, a stable amorphous–nanocrystalline structure forms which undergoes quantitative changes with a further increase of deformation.

5.7. Effect of the divisibility and direction of deformation

The deformation conditions (the type of stress state, temperature, rate, etc.) have a strong effect on the resultant structure and properties of materials. One of the factors of this type of effect is the divisibility of deformation, i.e., the number of cycles of external effects resulting in the required degree of plastic deformation [63]. In principle, all possible methods of deformation can be conventionally divided into two large groups: continuous and discontinuous. In the first case, the application of external load results in the continuous plastic deformation of the solid to the required degree of shaped changes retaining constant or increasing the external stress (for example, uniaxial tension, compression, bending, torsion, etc.). In the second case, the mechanical effect on the solid is of the short-term 'quantum' nature with the possibility of realising only one possible degree of plastic deformation in a single pass. The processes of plastic shape changes are in this case of the discontinuous nature and take place in several cycles (passes) (for example, rolling, pressing, forging, extrusion).

The number of cycles in the fractional deformation method has not been given so much attention as, for example, temperature or strain rate. However, the elastic relaxation processes, taking place during the 'break' between the individual passes even at room temperature are capable of exerting a strong effect on the defective structure and, consequently the properties of the material, subjected to a specific total degree of deformation. In particular, this applies to the cases in which the external effects result in very high deformation (severe plastic) of the material. When using the method of equal channel angular pressing – the most popular fractional method of producing SPD – the number of passes resulting in the very

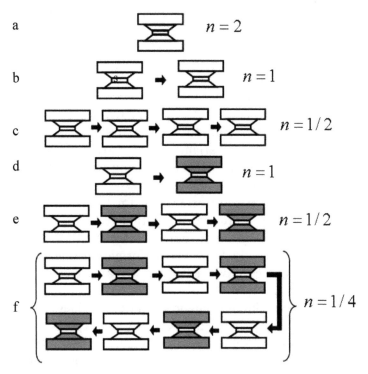

Fig. 5.36. Diagram of the experiments: a – continuous deformation, b – d – deformation of the type with different numbers of fractional deformation cycles, e, f – deformation of the type ↔ with different number of the divisible deformation cycles; n is the number of rotations in a single fractional deformation cycle. The white and grey colours indicate the rotation of the moving anvil in the direct and reversed directions, respectively.

high deformation has been mentioned only in a small number of articles [30]. In [64] it has been confirmed theoretically that the non-monotonic (in particular cyclic) loading of the material leads, with other conditions being equal, to less extensive refining of the grains and lower susceptibility to fracture in comparison with the process of quasi-monotonic loading.

In [33] it was attempted to evaluate by experiments the 'divisibility effect' in severe plastic deformation. The method of generating severe plastic deformation was based on the Bridgman chamber used for torsional deformation under high quasi-hydrostatic pressure. The material for the investigations was commercial purity iron, containing 0.05% C. The experiments were carried out using the following procedure (Fig. 5.36). The total deformation of the specimen in all cases corresponded to two full rotations of a moving anvil ($N = 2$) and in accordance with the equation (1) equalled $\varepsilon_N =$ 6.2. Deformation was carried out by several procedures (Fig. 5.36). The variant *a* was based on continuous deformation in a single partial pass *n* equal to 2 full rotations of the moving anvil ($N = 2$, $n = 2$). In variant *b* the specimen is deformed in two passes in the same direction of rotation of the moving anvil ($\rightarrow$). Each partial pass *n* equals 1 complete rotation ($N = 2$, $\rightarrow$, $n = 1$). Variant *c* corresponds to 4 partial passes in the same direction of rotation of the moving striker ($N = 2$, $n = \frac{1}{2}$). The variants *d* and *e* are identical with the variants *b* and *c* with the only difference being that the directions of rotation changed gradually to opposite ($\leftrightarrow$): $d - N = 2$, $\leftrightarrow$, $n = 1$; $e - N = 2$, $\leftrightarrow$, $n = \frac{1}{2}$. In addition to this, for the deformation with the gradually changing direction of rotation of the moving striker the variant *f* with an even larger number of partial passes ($N = 2$, $\leftrightarrow$, $n = \frac{1}{4}$) was formed. In fractional deformation, the duration of the 'breaks' between the individual cycles was approximately 20 min at room temperature.

Figure 5.37 shows the dependence of microhardness HV on the number of partial cycles *n* for the deformation experiments of the type '$\rightarrow$' (torsion in the same direction in each cycle) and the type '$\leftrightarrow$' (torsion with gradually changing direction in each consecutive cycle). It may be seen that the increase of the number of cycles at the constant total severe plastic deformation results in a large decrease of the value of HV, and the 'divisibility effect' is more distinctive for deformation of the type '$\leftrightarrow$' ($\Delta HV/HV > 10\%$).

Figure 5.38 shows the histograms of the size distribution of the grains and fragments for each variant of deformation obtained on the basis of the transmission electron microscopy data. A special feature

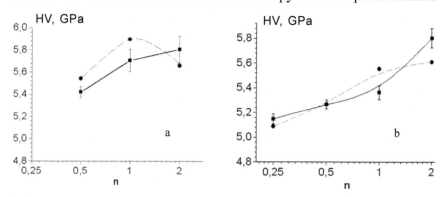

Fig. 5.37. Dependence HV (*n*): a – deformation of the type $\rightarrow$, b – deformation of the type $\leftrightarrow$, the solid and dotted lines – the experiments and theory, respectively.

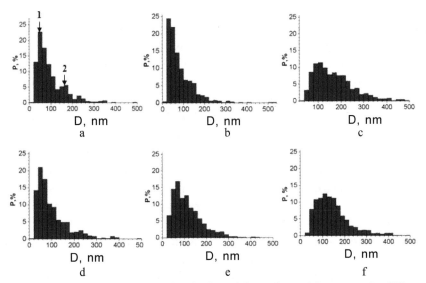

Fig. 5.38. Histograms of the size distribution of the grains and fragments for different deformation variants; the numbers indicate the numbers in Fig. 5.36.

of these distributions is the slightly lower value of the average size of the fragments in comparison with, for example, the data in the review [5] which is caused evidently by the commensurability of the dimensions of the fragments and the thickness of the investigated foil. In this case, the appropriate error is of the systematic nature and cannot influence in the present case the comparative characteristics of the size of the grains and fragments after different deformation treatments.

Analysis of the distributions in Fig. 5.38 shows the existence of two maxima (bimodal distribution), and the position and height of these maxima change depending on the type of deformation. This was already discussed previously in the examination of the structure of metals and solid solutions subjected to the SPD. These maxima are especially distinctive in continuous deformation ($N = 2$, $n = 2$) without intermediate arrests (maxima 1 and 2 in Fig. 5.38a). With increase of the number of deformation cycles the histograms are extensively transformed, but the bimodal form of the distributions remains, although it becomes slightly blurred. The 'divisibility effect', associated with the change of the nature of the histograms, is more evident in the deformation experiments of the type '↔'.

To describe the structure, formed in iron after SPD, the authors of [33] used the model of the 'two-phase mixture' mentioned previously by the authors of this book which consists of the recrystallised grains and deformation fragments. Figure 5.39 shows the typical histogram of the distribution of the mutual misorientations of the grains for different structural states produced by the EBSD method. There are two regions of distribution of the misorientations. One region corresponds to the low-angle and medium-angle grain boundaries with the misorientation angle of up to 35°, consisting mainly of the grains with the misorientation smaller than 3°. The second region of the histogram corresponds to the high-angle boundaries with the misorientation angle in the range 40–60°. It is rational to assume that the first group of the grains is of the deformation origin and formed as a result of the deformation fragmentation processes [22], whereas the second group of the grains formed as a result of the dynamic recrystallisation processes [29].

Subsequently, experiments were carried out for each deformation mode to determine the relative fractions of the deformation and recrystallised grains by calculating the relative areas below the Gaussian distributions. The calculated values of the fraction of the deformation fragments C_{fr} and the recrystallised grains C_{rec} are presented in Fig. 5.40 in dependence on the value of the discontinuous deformation cycle n for the experiments of the type '→' (Figure 5.40a) and the type '↔' (Fig. 5.40b). It may be seen that for the case '→' the 'divisibility effect' of severe plastic

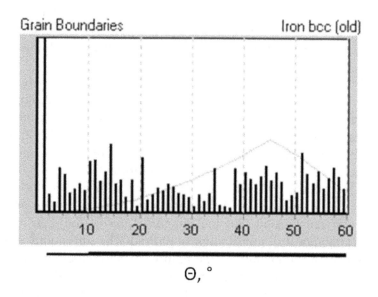

Fig. 5.39. The distribution of the misorientation of the grains and fragments for the structural state after deformation variant e (EBSD).

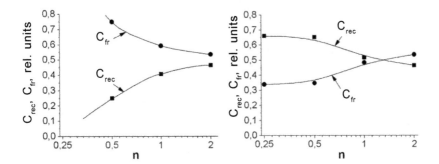

Fig. 5.40. The fraction of the recrystallised grains (C_{rec}) and fragments (C_{fr}) in dependence on the value of the fractional strain n of the type → (a) and ↔ (b).

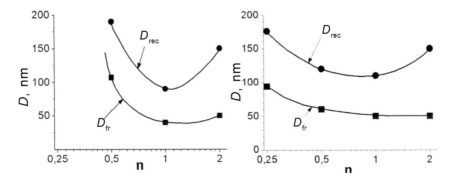

Fig. 5.41. The average size of grains (D_{rec}) and fragments (D_{fr}) in dependence on discontinuous deformation n of the type → (a) and ↔ (b).

deformation results in a large decrease of C_{rec} (to 0.25 at $n = \frac{1}{2}$) and in a corresponding increase of C_{fr} (up to 0.75 at $n = \frac{1}{2}$) which after continuous deformation ($n = 2$) were almost completely equal ($C_{rec} = C_{fr} = 0.5$). In the deformation experiments of the type '↔' this divisibility effect is completely different: the value C_{rec} tends to increase slightly with saturation at low values of n ($C_{rec} \rightarrow 0.65$), and the value C_{fr} tends to decrease with saturation at low values of n ($C_{fr} \rightarrow 0.35$).

Using the procedure of dividing experimental histograms into two components it is also possible to investigate the change of the average grain size corresponding to the fragments and recrystallised grains with the change of the number of cycles of severe plastic deformation. Figure 5.41 shows the dependence of the average grain size D of the fragments and recrystallised grains on n for the deformation experiments of the type '→' (Fig. 5.41a) and '↔' (Fig. 5.41b). The dependences $D(n)$ for both types of loading are similar: the average size of the fragments and recrystallised grains shows a characteristic minimum, corresponding to $n = 1$, increasing (more extensively for '→') with both the decrease and increase of n.

The experimental results show that the cyclic nature of the application of external loading has a strong effect on the structure and mechanical properties of the metal subjected to SPD. Consequently, it may be assumed that to describe correctly the deformation procedures used, it is important to indicate the number of cycles of application of the external load and the magnitude of plastic deformation in each load cycle. It may also be assumed that because of the general nature of the process, taking place at very high strains, the conclusions on the significance of the 'divisibility effect' are valid not only for torsional loading under pressure but also for equal channel angular pressing, screw extrusion, rolling, volume forging and other methods of severe plastic deformation in metallic materials.

The dependences $C(n)$ and $D(n)$, shown in Figs. 5.40 and 5.41 clearly demonstrate the 'divisibility effect' and can be explained unambiguously by the energy concept of the structural and phase transformations in severe plastic deformation [47] taking into account the fact that the increase of the divisibility of deformation and the change of the type of loading ('↔' instead of '→') decreases the elastic energy of the deformed material and complicates the processes of dynamic recrystallisation.

The dependence $HV(n)$ in Fig. 5.37 can be easily understood taking into account two evident assumptions:

1. The plastic flow resistance can be described by the Hall–Petch relationship $HV \sim D^{-\frac{1}{2}}$;

2. Both structural components (fragments and recrystallised grains) provide an additive contribution to hardening.

As a result of these functions

$$HV = HV_0 + k_{rec} \, C_{rec} \, (D_{rec})^{-1/2} + k_{fr} \, (1 - C_{rec})(D_{fr})^{-1/2} \quad (5.4)$$

where HV and HV_0 are the values of microhardness relating respectively to the alloys subjected to deformation in different conditions and to the crystalline matrix; C_{rec} and C_f are the volume fractions of the regions occupied respectively by the recrystallised grains and the fragments; k_{rec} and k_{fr} are the coefficients determining the 'permeability' of respectively the boundaries of the recrystallised grains and the fragment boundaries, and finally D_{rec} and D_{fr} are the average sizes of the recrystallised grains and the fragments.

Using the data presented in Figs. 5.40 and 5.41 and assuming that $k_{fr}/k_{rec} \cong 3$ using the equation (5.4) gives the analytical dependences shown in Fig. 5.37 which are in satisfactory agreement with the experimental dependences HV(m) for different deformation conditions. The fact that $k_{fr} > k_{rec}$ indicates that the less perfect or more distorted fragment boundaries are less permeable to plastic shear than the boundaries of the recrystallised grains. At the same time, the specific relationship between k_{fr} and k_{rec} is in fact a physically justified adjustable parameter between the theory and experiment.

Thus, the divisibility (the number of passes) and the method of deformation in SPD have, with other conditions being equal, a strong effect on the nature of the structure and mechanical properties of the polycrystals of commercial purity iron. The relationship between the regions of the structure corresponding to deformation fragments and recrystallised grains is determined within the framework of the energy concept of the structural and phase transformations in severe plastic deformation taking into account the fact that the increase of the divisibility of deformation and the change in the loading direction ('↔' instead of '→') reduce the level of the elastic energy of the deformed material and complicate the course of the dynamic recrystallisation processes. The change of the mechanical properties (microhardness) of the material with the variation of divisibility and the method of severe plastic deformation is quantitatively described on the basis of the assumptions that the plastic flow resistance can be described by the Hall–Petch relationship and both structural components (fragments and recrystallised grains) provide an additive contribution to hardening.

5.8. The principle of cyclicity in severe plastic deformation

The classic considerations regarding plastic deformation are based on the fact that the increase of the degree of deformation is accompanied by the buildup of dislocations defects. As the plastic strain increases, the number of defects which the deformed crystals should contain increases. The first exclusion from this rule formed when examining severe plastic deformation with the active participation of disclination modes: the fragments were characterised by thin boundaries and were almost completely free from dislocations. However, in transition

to the range of the SPD, as shown previously, there are dramatic
structural rearrangements as a result of additional dissipation channels
of elastic energy. The jump-like change of the structure and properties
in transition to SPD was reported by the authors of [65]. If attention
is given to a specific microvolume of the deformed specimen, then
the dynamic recrystallisation or amorphisation is followed by the
process of plastic deformation which appears to start 'from a blank
sheet' in the newly formed recrystallised grains or in the region of
the amorphous phase. Subsequently, the defects are built up again
in the investigated microvolume under the effect of the deformation
stresses and the process is repeated. A similar cyclic nature in SPD
was detected directly by, for example, the authors of [21].

Figure 5.42 shows the calculated curves of plastic flow taking
into account the existence of the dissipation channel; the curves were
obtained for materials with different mobility of the dislocations and
at different temperatures [25]. It may be seen that at low values of
the Peierls barrier σ_s and in the presence of the effective dissipation
channel the elastic flow curve is cyclic with the 'wavelength' $\Delta\varepsilon$. The
fact that similar flow curves are never recorded in the experiments
does not contradict this conclusion. The process of plastic flow in all
stages takes place, as is well-known, in highly non-homogeneous,
and different areas of the deformed material are in different stages
of their evolution.

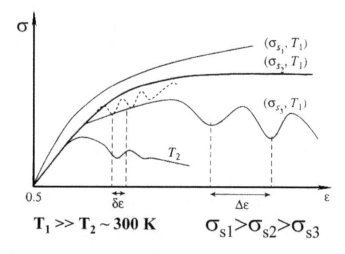

Fig. 5.42. Possible types of deformation curves in the region of SPD for materials
with different Peierls barrier $\sigma_{s1} > \sigma_{s2} > \sigma_{s3}$ at temperatures of $T_2 \gg T_1 = 300$ K.

We examine the 'roadmap' formed in SPD of pure metals and solid solutions [66]. The transfer of the deformed system to the stationary mode may take place by two possible mechanisms. In the first case, the processes of deformation fragmentation and dynamic recrystallisation are completely equalised and the system tends monotonically to its stationary state. In the second case, the complete equilibrium is not established, and the fragmentation processes initially outstrips the dynamic recrystallisation processes and, subsequently, on the other hand slightly lags behind them and this results in the vibrational nature of evolution of the deformed system.

If the relaxation of the excess elastic energy takes place locally irrespective of relaxation in other areas of the solid, then on the whole the relaxation will continuous and monotonic and the system will evolve monotonically to its equilibrium state by the first of the previously described scenarios. However, if relaxation is of the correlated avalanche-like form, when the transition to the equilibrium state of one region of the solid synchronises the transition to the equilibrium value of the parameters and variables in other stressed regions, the solid will efficiently relax and, this will be followed by the stage of the new build up of elastic energy for the next jump-like transition. If this process is quite distinctive, then on the whole the relaxation will be of the vibrational periodic nature in accordance with the second scenario mentioned previously.

To describe similar vibrational transitions on the basis of non-equilibrium evolution thermodynamics, the authors of [67] propose a model of evolution equations with inertia (with a memory). In the first approximation, such a model can be used to describe the phenomena associated with the periodicity of the processes of deformation fragmentation and recrystallisation of the grains. The system of the evolution equations taking into account the contribution of the boundaries of the dislocation fragments and the boundaries of the recrystallised grains for SPD has the form:

$$\frac{\partial h_i}{\partial t} = \gamma_i \left(\frac{1}{T_i} \int_{t-T_i}^{t} f_i(t') \frac{\partial u}{\partial h_j} dt' - \varphi_i \right), \tag{5.5}$$

where h_i is the density of defects. Index $i = g$ relates to the grain boundaries, index $i = D$ to the boundaries of the dislocation fragments; φ_i is the energy of the appropriate defect; γ_i is the kinetic coefficient; $f_i(t')$ is the core of the integral transformation which describes the degree (rate) of 'forgetting' previous states; T_i is the time period during which the system completely 'forgets' the previous states.

If the function $f_i(t')$ is equal to unity in the range $0 \le t' \le T_i$, and to 0 outside this range, the integral represents the mean arithmetic value of the generalised thermodynamic force which acts throughout this time period. The interval T_i in this case is an additional controlling parameter of the problem. In the case of SPD, h_g is the bulk density of the total surface of the grain boundaries, φ_g is the surface energy density of the grain boundaries in the stationary state, h_D is the bulk density of the total surface of the dislocation fragments; φ_D is the surface energy density of the dislocation fragments in the stationary state.

The dependence of internal energy on the parameters of the problem in the general case has the form

$$u\left(h_g, h_s\right) = u_0 + \sum_{m=g,D} \left(\varphi_0 h_m - \frac{1}{2}\varphi_{1m} h_m^2 + \frac{1}{3}\varphi_{2m} h_m^3 - \frac{1}{4}\varphi_{3m} h_m^4 \right) + \varphi_{gD} h_g h_D, \quad (5.6)$$

where $u0$, φ_{km} (index $k = 0, 1, 2, 3$), φ_{gD} are the polynomial coefficients, which depend on the entropy s and elastic strain ε_{ij}^e, as on the controlling parameters

$$u_0 = \frac{1}{2}\lambda\left(\varepsilon_{ii}^e\right)^2 + \mu\left(\varepsilon_{ij}^e\right)^2 + \beta s^2,$$

$$\varphi_{0m} = \varphi_{0m}^* + g_m \varepsilon_{ii}^e + \left(\frac{1}{2}\bar{\lambda}_m \left(\varepsilon_{ii}^e\right)^2 + \bar{\mu}_m \left(\varepsilon_{ij}^e\right)^2 \right) - \beta_m s + \beta_{gm} s \varepsilon_{ij}^e, \quad (5.7a)$$

$$\varphi_{1m} = \varphi_{1m}^* - 2e_m \varepsilon_{ii}^e. \quad (5.7b)$$

The kinetic (evolution) equations (5.7) without taking the history of the process in the explicit form into account and retaining the quadratic contributions for the dislocation fragments have the form

$$\frac{\partial h_g}{\partial t} = \gamma_g \left(\varphi_{0g} - \varphi_{1g} h_g + \varphi_{3g} h_g^2 + \varphi_{4g} h_g^3 - \varphi_g + \varphi_{gD} h_d \right) \quad (5.8a)$$

$$\frac{\partial h_D}{\partial t} = \gamma_D \left(\varphi_{0D} - \varphi_{1D} h_D + \varphi_{gD} h_g \right). \quad (5.8b)$$

The positive terms in this expression are responsible for the generation of deformation fragments, the negative terms describe

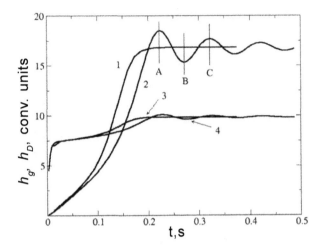

Fig. 5.43. Kinetic curves of formation of defects of different type: 1, 2) the density of the grain boundaries; 3, 4) the density of deformation fragments. 1, 3) disregarding the 'prior history', 2, 4) taking the 'prior history' into account, A, B, C – extreme values.

the reverse processes of annihilation of defects, in particular, as a result of dynamic recrystallisation by one of the mechanisms of 'diffusionless' dynamic recrystallisation.

The following surface of the parameters and coefficients were used for the calculations: $\lambda = \mu = 2.08 \cdot 10^{10}$ Pa; $\varphi_{0D}^* = 5 \cdot 10^{-9}$ J·m^{-1}; $\varphi_{1D}^* = 10^{-24}$ J·m; $g_D = 2 \cdot 10^{-8}$ J·m^{-1}; $\overline{\mu}_D = 3.3 \ 10^{-4}$ J·m^{-1}; $e_D = 6 \cdot 10^{-23}$ J·m; $\varphi_{0g}^* = 0.4$ J·m^{-2}, $\varphi_{1g}^* = 3.10 \cdot 10^{-6}$ J·m^{-1}, $\varphi_{2g}^* = 5.6 \cdot 10^{-6}$ J; $\varphi_{3g}^* = 3 \cdot 10^{-20}$ J·m; $g_g = 12$ J·m^{-2}, $\overline{\lambda} = 2.5 \cdot 10^5$ J·m^{-2}; $\overline{\mu} = 6 \cdot 10^5$ J·m^{-2}; $e_g = 3.6 \cdot 10^{-4}$ J·m^{-1}, $\varphi_{gD} = 10^{-16}$ J.

These parameters are described by the kinetic curves 1 and 3 in Figure 5.43.

The kinetic curves for the defect densities without considering the 'history' of the SPD process are described by the positively changing functions, approaching monotonically the stationary value. This indicates that the processes of generation of the deformation fragments and dynamic recrystallisation are balanced throughout the entire duration of SPD. The generation of fragments increases the level of non-equilibrium which activates to the same extent proportionately the processes of dynamic recrystallisation, and both processes take place synchronously in time.

To take the vibrational evolution into account, it is assumed that $T_g = 0.43$ s, $T_D = 0.027$ s (curves 2 and 4 in Fig. 5.43). It may

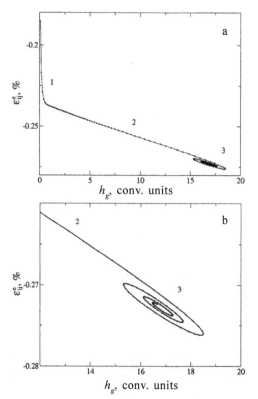

Fig. 5.44. The hardening–softening curves in the cyclic process; a – the total curve, b – the curve on the larger scale.

be seen that in this case at the same parameters of the model the evolution rate rapidly decreases, and the system reaches the same values of the defect density at later stages. In addition to this, in the first stage the recrystallisation processes lag behind the processes of generation of fragments and, consequently, the grains are refined to higher degrees than the degrees which could be reached in the stationary state. After reaching some maximum threshold level of the defect density, on the other hand, the rate of the processes of dynamic recrystallisation increases and these processes start to dominate all the processes of deformation fragmentation (section AB).

Figure 5.44 shows the correlation curve of the dependence of the limiting elastic deformation on the density of the grain boundaries taking the 'prior history' into account. In fact, the curve reproduces the strain hardening curve. Section 1 can be approximated by Hall–Petch relationship, section 2 corresponds to the linear hardening

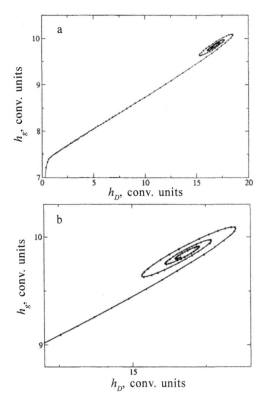

Fig. 5.45. Correlation between the density of the grain boundaries and the fragments: a – the total curve, b – the fragment of the curve on the larger scale.

law. The sections 1 and 2 coincide almost completely with the identical curves without taking the 'prior history' into account, and qualitatively correspond to the hardening laws observed in the experiments (for example in [68]). Section 3 is associated with the cyclic evolution stage. The information on the time dynamics of evolution of the system can be obtained from the points on the graph which were plotted over the identical time periods. In stage wonder rate of the processes are relatively high and, therefore, the process rapidly slows down and then again accelerates. In the final stage, the process also slows down and gradually approaches the stationary value.

It is interesting to examine the correlation dependence of the density of the boundaries of deformation fragments and the boundaries of the recrystallised grains (Fig. 5.45). The qualitative form of the curve is the same as that examined previously. It should be noted that the straight section in this case indicates that the rate of formation of the fragment boundaries and the grain boundaries does not change or at least changes in the same manner in the section of the highest rate of SPD (the section from 0 to point A in Figure 5.43).

On the basis of the theoretical examination results, carried out in [66] it may be assumed that to describe the experimentally observed variations of the mean grain size (or the density of the grain boundaries) in SPD we can use efficiently the concept of non-equilibrium evolution thermal parameters with nature (with a memory). Taking into account the memory effect (inertia) in deformation structure formation reduces the main relationships of the form characteristic for wave equations with damping. In the approximation of the one-modal distribution with respect to the size of the grains the deformation system gradually (carrying out damping oscillations) transfer to the stationary value of the structural parameters. It may be assumed that SPD is accompanied by cyclic changes of not all of the structural parameters but also the strength properties of the material changing the form of the final stage of the hardening curve.

5.9. Conclusions

Summarising these chapter, it is important to note the main special features of the process of high (severe plastic) deformation which greatly differ from other stages of plastic yielding.

– The process of SPD must be accompanied by the formation of additional (in addition to elastic deformation of fracture) channels of dissipation of elastic energy. The structural changes in SPD are characterised by the cyclic form;

– the specific route of the structural rearrangement in SPD is determined by a number of factors (temperature, the size of the Peierls barrier of the dislocations and their capability for diffusion rearrangement, the differences in the free energies of the crystalline and amorphous state);

– the occurrence of SPD does not guarantee at all the formation of the nanocrystalline state with the crystals smaller than 100 nm, separated by high-angle or interphase boundaries. For example, this is not possible in the pure metals with high dislocation mobility. An important factor of the formation of the nanostructures in the SPD is the occurrence of phase transformations of the martensitic and diffusion type, and also transition to the amorphous state. Stimulating the phase transformations by varying the temperature and chemical composition of the materials, it is possible to produce nanostructures of different types;

– the additional effective channels of dissipation of elastic energy in SPD are: dynamic recrystallisation, dislocation–disclination accommodation, phase transformations (including transition to the amorphous state) and the release of latent heat of the deformation origin. At the (macroplastic) stresses the elastic energy gradually builds up and powerful dissipation processes start to operate only in the SPD stage;

– it is possible to determine accurately the critical deformation region in which the macroplastic deformation changes to severe plastic. The boundary region of implementation of the SPD is determined accurately by the start of activation of one of the possible additional dissipation channels. From the morphological viewpoint, the SPD stage can be recorded by, for example, appearance in the structural dynamics of the recrystallised grains or microregions of the amorphous phase. If several previously mentioned dissipation channels operate at the same time, the boundary value of deformation corresponds to the activation of the dissipation channel corresponding to lower plastic strain values;

– the SPD stage is usually preceded by the stage of development of plastic deformation characterised by the active occurrence of the processes of deformation fragmentation with the formation of regions relatively free from defects and separated by boundaries with a wide spectrum of the crystallographic misorientations, including low, medium and high angular values.

References

1.   Fortov V.E., Extreme states of matter, Moscow, Fizmatlit, 2009.
2.   Valiev R.Z., et al., Bulk Nanostructural Materials, Wiley & Sons, 2014.
3.   Bridgman P.W., Studies of large plastic deformation and fracture. The influence of high hydrostatic pressure on the mechanical properties materials. (Translated from English, ed. L.F. Vereshchagin), Moscow, LIBROKOM 2010.
4.   Bridgman P.W., *J. Appl. Phys.*, 1946, V. 17, 692.
5.   Zhilyaev A.P., Langdon T.G., *Prog. Mater. Sci.*, 2008, V. 53, 893–979.
6.   Segal V.M., et al., The processes of plastic metal structure formation, Minsk, Navuka i tehnika, 1994..
7.   Valiev R.Z., Alexandrov I.V., Nanostructured materials obtained by severe plastic deformation, Moscow, Logos, 2000.
8.   Utyashev F.Z., Raab G.I., Deformation methods of obtaining and processing ultra-

fine-grained and nanostructured materials, Ufa, Guillem, NIK Bashk. Encyc., 2013.

9.   Shtremel' M.A., The strength of the alloys, Part 2, Moscow Institute of Steel and Alloys, 1997.

10.   Degtyarev M.V. Stages of development of ultrafine structures in iron and structural steels at high pressure deformation (Thesis for the degree of Doctor of Technical Sciences), Ekaterinburg, 2005.

11.   Valiev R.Z., et al., *JOM*, 2006, V. 58, No. 4, 33–39.

12.   Large English-Russian Dictionary, Moscow, Russkii yazyk, 1988, Vol. 2, 271.

13.   Glezer A.M., *Izv. RAN, Ser. Fiz.*, 2007, V. 71, No. 12, 1764–1772.

14.   Golovin Yu.I., The universal principles of natural science. Tambov, TSU 2002.

15.   Microplasticity. Collection of articles edited by Mc Mahon (trans. from English) Moscow, Metallurgiya, 1972.

16.   Glezer A.M., *Usp. Fiz. Nauk*, 2012, V. 182, No. 5, 559–566.

17.   Zhilyaev A.P., Langdon T.G., *Progr. in Mater. Science*, 2008, V. 53, 893–979.

18.   Koneva N.A., et al., *Metallofizika*, 1991, No. 10, 49–58.

19.   Gapontsev V.L., Kondratyev V.V., *Dokl. Akad. Nauk*, 2002, V. 385, No. 5.684–687.

20.   Chuvil'deev V.N., Non-equilibrium grain boundaries in metals. Theory and application, Moscow, Fizmatlit, 2004.

21.   Bykov V.M., et al., *Fiz. Met. Metalloved.*, 1978, V. 45, No. 1, 163–169.

22.   Rybin V.V., Large plastic deformation and fracture of metals, Moscow, Metallurgiya, 1986..

23.   Svirina Yu.V., Perevezentsev V.N., *Deformatsiya i razrushenie metallov*, 2013, No.7. 2–6.

24.   Glezer A.M., Metlov L.S, *Fiz. Tverd. Tela,* 2010, V. 52, No. 6, 1090–1097.

25.   Pozdnyakov V.A., Glezer A.M., *Izv. RAN, Ser. Fiz.*, 2004, V. 68, No. 10, 1449–1455.

26.   Gorelik S.S., et al., Recrystallization of metals and alloys, Moscow Institute of Steel and Alloys, 2005.

27.   Kassner M.E., Barrabes S.R., *Mater. Sci. Eng.*, 2005, V. A 410–411, 152–155.

28.   Sakai T.,et al., *Acta Mater.*, 2009, V. 57, 153–162.

29.   Humphreys F.J., Hatherly M., Recrystallization and related annealing phenomena, Amsterdam, Elsevier, 2004..

30.   Estrin Y., Vinogradov A., *Acta Mater.*, 2013, V. 61, 782–817.

31.   Glezer A.M., et al., *DAN*, 2014, V. 457, No. 5, 535–538.

32.   Blinova E.N., Glezer A.M., Materialovedenie, 2005, No. 5, 32–39.

33.   Glezer A.M. et al., *Deformatsiya i razrushenie metallov*, 2014, No. 4, 15–20.

34.   Andrievsky R.A., Glezer A.M., *Usp. Fiz. Nauk,* 2009, V. 179, No. 4, 337–358.

35.   Nalimov V.V., Experiment theory, Moscow, Nauka, 1974.

36.   Tat'yanin E.V., et al., *Fiz. Met. Metalloved.*, 1986, V. 62, No.1, 133–137.

37.   Brailovski V., et al., *Materials Transaction JIM*, 2006, V. 47, No. 3, 795–804.

38.   Chen H., et al., *Lett. Nature,* 1994, V. 367, No. 2, 541–543.

39.   Gunderov D.V., et al., *Deformatsiya i razrushenie metallov*, 2006, No. 4, 22–25.

40.   Glezer A.M., et al., *Mater. Sci. Forum*, 2008, V. 584–586, 227–230.

41.   Glezer A.M., et al., *Izv. RAN, Ser. Fiz.,* 2009, V. 73, No. 9.,S. 1302–1309.

42.   Glezer A.M., et al., Mechanical behavior of amorphous alloys, Novokuznetsk, SGIY, 2006..

43.   Glezer A.M., et al., *Izv. RAN, Ser. Fiz.,* 2012, V. 76, No. 1, 63–70.

44.   Sundeev R.V., et al. *Deformatsiya i razrushenie metallov*, 2013, No. 5, 2–9.

45.   Glezer A.M., et al., *Izv. RAN, Ser. Fiz.,* 2013, V. 77, No. 11, 1687–1692.

46.   Sundeev R.V., et al., *Mater. Lett.*, 2014, V. 133, 32–34.

47.   Glezer A.M., Sundeev R.V., *Mater. Lett.*, 2015, V. 139, 455–457.

48.   Donovan P.E., Stobbs W.M., *Acta Met.*, 1981, V. 29, No. 6, 1419–1424.
49.   Kovneristyi J.K., et al., *Deformatsiya i razrushenie metallov*, 2008, No. 1, 35–41.
50.   Glezer A.M., et al., *Russian Metallurgy (Metally)*, 2013, No 4, 235–244.
51.   Grjaznov V.G., et al., *Pis'ma ZhTF*, 1989. T. 15. No. 2. S. 1256–1261.
52.   Glezer A.M. et al., *DAN*, 2011, V. 440, No. 1, 39–41.
53.   Pushin V.G., et al., in: The structure, phase transformations and properties, Ekaterin-burg, IMP UB RAS, 2006.
54.   Tat'iyanin E.V., et al., *Fiz. Tverd. Tela*, 1997, V. 39, 1237–1243.
55.   Prokoshkin S.D., et al., *Acta Mater.* 2005. V. 53, 2703–2714
56.   Zel'dovich V.I., et al., *Fiz. Tverd. Tela*, 2005, V. 99, 90–98.
57.   Prokoshkin S.D., et al. *Fiz. Met. Metalloved.*, 2010, V. 110, No. 3, 305–320.
58.   Gunder D.V., Electronic scientific journal: *Issledovano v Rossii*, 2006.
59.   Pushin V.G., et al., *Fiz. Met. Metalloved.*, 1997, V. 83, No. 6, 149–156.
60.   Nosov G.I., et al., *Kristallografiya*, 2009, V. 54, No. 6, 1111–1119.
61.   Sherif El-Eskandarany M., et al., *Acta Met.*, 2002, V. 50, P. 1113–1123.
62.   Glezer A.M., et al., *Izv. RAN. Ser. Fiz.*, 2010, V. 74, No. 11, 1576–1582.
63.   Honeycome R.W., Plastic deformation of metals (Translated from English.), Mos-cow, Mir, 1972.
64.   Beygelzimer Y., *Mechanics of Materials*, 2005, V. 37, 753–767.
65.   Firstov S.A., et al., *Izv. VUZ, Fizika*, 2002, No. 3, 41–48.
66.   Metlov L.S., et al., *Deformatsiya i razrushenie metallov*, 2014, No. 5. 8–13.
67.   Metlov L.S., *Visnik Dontesk. Univ., Ser. A, Prirod. Nauki*, 2009, V. 2, 136–152.
68.   Podrezov Yu.N., Firstov S.A., *Fiz. Tekh. Vysokoho Davleniya*, 2006, V. 16, No. 4, 37–48.

6

Effect of ion implantation on structural state, phase composition and the strength of modified metal surfaces

6.1. Introduction

The methods of treatment of materials with the beams of metallic ions are present one of the most extensively developed direction of synthesis of new materials [1–12]. At present, all experimentally justified and developed models of phase formation in ion implantation have been developed for coarse-crystalline metals [1–28]. However, the decrease of the grain size and, correspondingly, increase of the density of intergranular boundaries and triple junctions may result in a large increase of the rate of diffusion processes, alloying, mixing, formation of secondary phases and defects [9], and also the appearance of new, including intermetallic phases. Of considerable interest is the ion synthesis of different phases formed in the surface layers of materials based on titanium and aluminium. This is explained by the fact that the intermetallic phases of the Ti–Al system are characterised by high mechanical strength, hardness, wear resistance and corrosion resistance [28, 29] and usually have high melting points [30].

This chapters examines as an example special features of the processes of structure formation and formation of mechanical properties in ion implantation on the surface of metallic titanium – a promising material for application in practice. The formation of nanoparticles of intermetallic phases in the structure of a titanium-based alloy in the ion implantation conditions should result in considerable hardening of this material determined by both the dispersion hardening and the formation of internal stress fields [31]. Therefore, these materials can be used as alternatives of coarse crystalline titanium alloys for medical [32, 36] and technical applications used at present [34, 35]. Nevertheless, ion implantation of aluminium in titanium carried out to improve the surface properties of the latter has not been studied sufficiently. For example, in [8] investigations were carried out into the formation of the surface layer of titanium containing intermetallic compounds (in particular Ti_3Al) as a result of high-dose ion implantation of aluminium ions. The studies [11, 12] are also concerned partially with this problem. However, analysis of the published experimental data for ion implantation in the Ti–Al system shows that the implantation of ions of titanium and aluminium has been studied in most cases. The implantation of aluminium ions in the titanium matrix has been studied only in a small number of investigations. In addition, the problems such as the change of the microstructure and phase composition in the thickness of the ion-implanted layer of the target have not been studied at all.

Of special interest from the viewpoint of the processes of phase formation in the ion implantation conditions are metallic polycrystals containing the matrix grains in a wide size range from 100 nm to 10–20 μm. This is due to the fact that a change of the grain size results in a change of the defective structure of the grains and the boundary zone, the structure of the grain boundaries, internal stresses, the deformation mechanisms and dislocation sources [11, 12]. Taking into account the physical mechanisms of plastic deformation, the polycrystalline aggregates can be divided into two groups on the basis of the grain size [36, 37]. The mesolevel includes the polycrystals with the grain size of 1 μm–10 mm and is characterised by the conventional behaviour according to the Hall–Petch relationship. The transition to the microlevel (the grain size 0.1–1 μm) initially results in considerable hardening of the polycrystalline aggregates and further grain refining results, in addition to the hardening in accordance with the Hall–Petch relationship, in the dependence of the parameter K on the grain size and in a decrease of the extent of grain boundary hardening [36–39]. The occurrence of many physical processes on the mesolevel and microlevel differs [40, 41]. For example, there is a large difference in the relationships governing the buildup of dislocations on the mesolevel and microlevel with respect to the grain size: with increasing grain size on the mesolevel the dislocation density decreases, and on the microlevel it increases [40, 41]. In the ultrafine-grained polycrystals the dislocations play a significant role and in these materials the role of the grain boundaries, ternary and quaternary junctions becomes important [42]. On the microlevel of the grain size a significant role is played by the contribution of the inhibition of the grain boundaries to the flow stress [40].

Therefore, it is very important to investigate the processes of phase formation which take place during the implantation of titanium materials with aluminium alloys, with the materials being in different actual states (on the mesolevels and microlevels). The subject of the present chapter is the presentation and analysis of the results of investigation of the evolution of the structural–phase state, the grain structure and the parameters of the dislocation substructure of titanium after the effect of aluminium ions in dependence on the structural state of the target. The materials in this chapter generalise the experimental results, obtained previously by the authors of the book in [30–23, 26, 27, 43–64].

The investigations of the effect of ion implantation on the structural characteristics of titanium were carried out on three types of titanium targets: 1) $Ti_{0.3}$ – titanium specimens in the ultrafine-grained (UFG) state, the average size of the structural elements 0.3 μm; 2) $Ti_{1.5}$ – titanium specimens in the fine-grained (FG) state, the mean grain size 1.5 μm; 3) Ti_{17} – titanium specimens in the mesopolycrystalline state (MPC), the average grain size 17 μm. It should be noted that here we use the classification of the polycrystals on the basis of the grain size according to the data in [36–38, 40, 42, 65–67] and also the data of the authors of the present book [27].

The ion implantation of the titanium materials was carried out in a MEVVA-V.RU ion source [68, 69] at a temperature of 623 K, the accelerating voltage 50 kV, the ion beam current density of 6.5 nA/cm², the distance from the ion-optical system 60 cm, implantation time 5.25 h, irradiation dose $1 \cdot 10^{18}$ ion/cm².

6.2. Effect of ion implantation on the structure of titanium alloys

As mentioned previously, investigations were carried out on three types of titanium targets with different grain sizes: 1) in the ultrafine grained state with the average size of the structural elements of 0.3 μm (Ti$_{0.3}$); 2) in the fine-grained state with the average grain size of 1.5 μm (Ti$_{1.5}$), and 3) in the mesopolycrystalline state with the average grain size of 17 μm (Ti$_{17}$). The typical images of the grain structure of the initial samples (prior to implantation) obtained by different methods are presented in Fig. 6.1. Since the investigations were carried out on targets with greatly differing average grain sizes, the grain structure was investigated by different methods. In particular, the ultrafine-grained state was studied by transmission electron microscopy on thin foils, the fine grain size and mesopolycrystalline grain size – by the methods of optical and scanning electron microscopy on sections etched to the intra-granular structure and the grain boundaries. As indicated by Fig. 6.1a, the grain structure of the Ti$_{0.3}$ alloy consists of the grains of the anisotropic form, with the average grain size of approximately 0.3 μm. Attention should be given to the fact that when describing the structure discussed in this chapter, a significant role is played by the density of the grain boundaries. In the Ti$_{0.3}$ alloy the boundary density is determined mostly by the transverse grain size. The longitudinal grain size introduces no changes in the value of the density of the grain boundaries. Therefore, in this work, the condition of the Ti$_{0.3}$ alloy is related to the ultrafine-grained state and it is assumed that the average grain size of this alloy is 0.3 μm.

The experimental results obtained by the authors of the present book show that the structure of the ultrafine-grained titanium (Ti$_{0.3}$) contains elongated elements (grains) with a distinctive texture. It should be mentioned that the anisotropic form of the grains is determined by the methods of preparation of the specimen: multiple uniaxial pressing (ABC-pressing) followed by multipass rolling in groove rolls. The transverse grain size in Ti$_{0.3}$ is in the size range 0.15–0.65 μm (Fig. 6.1a). Approximately 75% of the volume in the structure is occupied by the grains with the size smaller than 0.4 μm. The distribution function is of the single-modal time. The average grain size is $<d>$ = 0.34±0.06 μm. The maximum of the distribution function is located in the vicinity of the average value. The longitudinal grain size is in the range 1.5–2.0 μm (Fig. 6.1a). The distribution function is also of the single-modal type with a maximum in the vicinity of the average value. The main longitudinal size is $<L>$ = 2.17±0.47 μm. The anisotropy coefficient $k = L/d$ was determined, and its value was equal to 6.4.

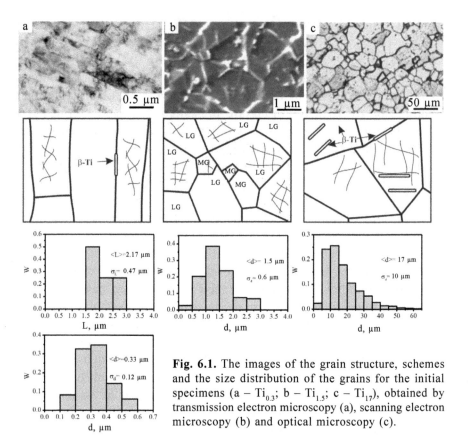

Fig. 6.1. The images of the grain structure, schemes and the size distribution of the grains for the initial specimens (a – Ti$_{0.3}$; b – Ti$_{1.5}$; c – Ti$_{17}$), obtained by transmission electron microscopy (a), scanning electron microscopy (b) and optical microscopy (c).

The structure of the fine-grained titanium (Ti$_{1.5}$) is characterised by the conventional polyhedral grain structure (Fig. 6.1b). There are two types of grains: large and fine. The fine grains are distributed either at triple junctions of the large grains or in the form of small groups. The volume fraction of the fine grains is approximately 10%. The size distribution of the grains is of the single-modal type (Fig. 6.1b). The maximum of the distribution function is found in the vicinity of the mean value of the grain size.

The structure of the mesopolycrystalline titanium (Ti$_{70}$) also contains the conventional polyhedral grain structure (Fig. 6.1c). The grain size changes in the range from 3 µm to 58 µm (Fig. 6.1c). However, the size of the majority of the grains (~70% of the total number) corresponds to the range 8–18 µm. The mean grain size is $d = 17\pm5$ µm. The size distribution of the grains is of the single-modal type. The maximum of the distribution function is found in the vicinity of the average value.

The experimental results show that the ion effect results in a large change of the grain state of titanium and is determined by the grain size of the target. For example, the investigation of the grain structure of the implanted layer in the ultrafine-grained matrix ($Ti_{0.3}$) in the thickness shows that the implantation of aluminium in titanium results in extensive refining of the grains of α-titanium (Fig. 6.2a, b). At a distance of 0–200 nm from the irradiated surface (the region I – the surface of the ion-doped layer) the average longitudinal grain size was almost halved and equalled 1.27 ± 0.51 μm (Fig. 6.2i, curve 1). The size range decreases (0.5–0 μm). The form of the size distribution of the grains changes and becomes multi-modal (Fig. 6.2i, curve 2). The size range decreases (0.05–0.40 μm). The distribution function also becomes multi-modal (Fig. 6.2c).

With a further increase from the irradiated surface to a distance of 250–450 nm (region II – the central part of the ion-doped layer) both the longitudinal and transverse grain sizes continue to decrease. The size ranges also continue to become smaller (Fig. 6.2i, curves and 1, 2). The distribution functions again become single-modal (Fig. 6.2f, g).

Implantation of $Ti_{0.3}$ is accompanied by a decrease of the coefficient of anisotropy of the grains (K), and with increase of the distance from the surface the value of K decreases even further (Fig. 6.2i, curve 3). This means that after implantation the grains become more isotropic. This fact becomes even more pronounced with increase of the distance from the implanted surface. Thus, the implantation of aluminium in titanium leads mostly to the formation of transverse boundaries and this is followed by the formation of longitudinal boundaries.

Investigations of the fine-grained titanium ($Ti_{1.5}$), carried out by scanning and transmission electron microscopy showed that implantation did not change the qualitative pattern of the grain structure of the material. As in the initial state, the entire implanted layer shows the conventional polyhedral grain structure (Fig. 3a, d). After implantation of aluminium in titanium, as in the initial state, the material contains two types of grains of α-titanium. The mutual distribution of the grains is the same as prior to implantation. The size distribution function in the entire implanted layer remains single-modal (Fig. 6c, f). The distribution function maxima for each region are situated in the vicinity of their average values. At the same time, it may be seen that with increase of the distance from the irradiated surface the average grain size changes. For example, at a distance

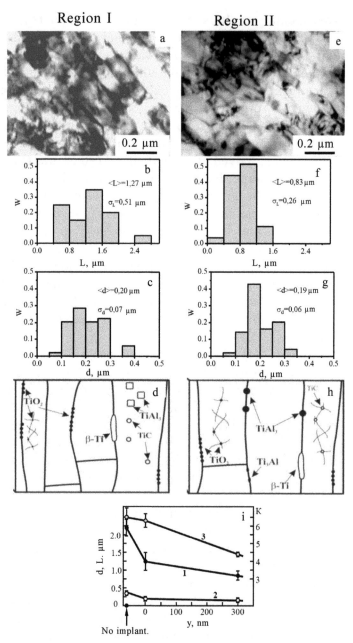

Fig. 6.2. Transmission electron microscope images (a, e), the size distribution histograms of the grains (b, f – longitudinal, and c, g – transverse dimensions) prior to (a–c) and schematic representation of the localisation of secondary phases (d, h) after ion implantation in the region I (a–d) and II (e–h) of the implanted layer of ultrafine-grained titanium ($Ti_{0.3}$) and the dependence of the average grain size (i) (1 – longitudinal L, 2 – transverse d) and the anisotropy coefficient K (3) prior to and after ion implantation.

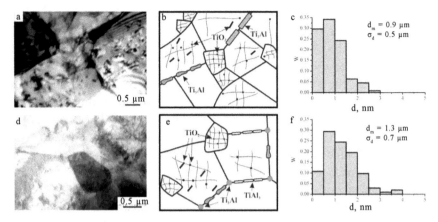

Fig. 6.3. Transmission electron microscope images of the surface layers of titanium (a, d) and the schemes of localisation of the titanium grains and the grains of the second phase in relation to the Ti grains (b, e): a, b, c – the ion-doped layer in the region I; d, e, f – the modified layer in the region II.

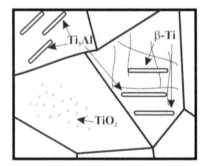

Fig. 6.4. Localisation of the phases formed in the implantation conditions.

of 0–200 nm from the irradiated surface (the region I), the average size of the grains of α-Ti is 0.9$\pm$0.5 μm. In the central region (the region II) of the implanted layer (at a distance of 250–450 nm from the surface) the average grain size is almost the same as prior to implantation.

The investigations of mesopolycrystalline titanium (Ti_{17}) show that the implantation of aluminium in titanium also did not change the qualitative pattern of the grain structure but resulted in grain refining (Fig. 6.4, regions I and II).

6.3. Distribution of implanted elements in the thickness of the implanted layer of titanium alloys

According to the results of Auger spectroscopy the maximum

concentration of the implanted impurity of aluminium in the implanted layer of the ultrafine-grained titanium ($Ti_{0.3}$) was not greater than 30 at.% (Fig. 6.5a, curve 1). The thickness of the entire implanted layer did not exceed ~600±50 nm. In the implantation in titanium with the average grain size of 1.5 μm (fine-grained titanium) there was a large increase of the concentration of the interstitial impurities (up to 60 at.%) with a small increase of the thickness of the doped layer (~650±15 nm) (Fig. 6.5a, curve 2). The surface layer of MPC titanium (Ti_{17}), implanted with aluminium, was characterised by the thickness of the modified layer of up to ~850±50 nm, and the concentration of the implanted impurity of up to 70 at.% (Fig. 6.5a, curve 3).

The Auger spectroscopy results also show that in addition to aluminium, the surface layers of all titanium targets contained

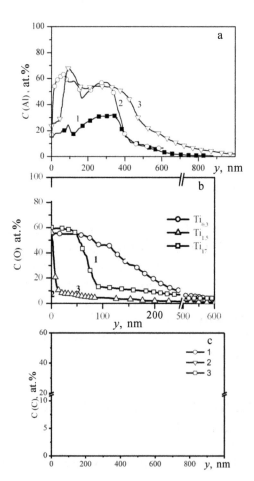

Fig. 6.5. The concentration of the Auger spectra of aluminium (a), oxygen (b) and carbon (e) in the surface layers of titanium after ion implantation: 1 – ultrafine grained titanium ($Ti_{0.3}$); 2 – fine-grained titanium ($Ti_{1.5}$); 3 – mesopolycrystalline titanium (Ti_{17}).

oxygen and carbon (Fig. 6.5b, c), implanted from the residual atmosphere of the vacuum system. The presence of oxygen and carbon, accompanying ion implantation, was noted previously in [70]. The maximum oxygen concentration (up to 50 at.%) was found in the ultrafine-grained titanium ($Ti_{0.3}$) as a result of easy conditions for diffusion. The carbon concentration did not exceed 10 at.%, and in the fine-grained titanium ($Ti_{1.5}$) and in MPC titanium (Ti_{17}) the carbon was not present. The observed concentrations of the alloying elements in the surface modified layers (Fig. 6.5b, c) were sufficient for the formation of secondary phases in the ion-doped layers.

The results of x-ray diffraction analysis confirm the data obtained in Auger spectroscopy. Table 6.1 shows the values of the parameters of the crystal lattice of the α-solid solution for the ultrafine-grained, fine-grained and mesopolycrystalline titanium. The table shows that the maximum crystal lattice parameters are obtained for the ultrafine-grained titanium ($Ti_{0.3}$).

These results may indicate that the solid solution of α-titanium contains a small amount of oxygen. The dependence of the crystal lattice parameter of the α-solid solution of oxygen in titanium on the composition shows that the solid solution has the form $TiO_{0.1}$ [71]. Increase of the grain size of the titanium target results in a decrease of the crystal lattice parameter and, correspondingly, a decrease of the oxygen content of the solid solution of oxygen in titanium (formula $TiO_{0.05}$). Ion implantation results in a decrease of the crystal lattice parameter because the surface of the specimen is heated during the process and this results in the 'cleaning' of the titanium lattice to

Table 6.1. Crystal lattice parameters of the investigated materials

Alloy		a, nm	c, nm	c/a
UFG-titanium ($Ti_{0.3}$)	initial	0.29601	0.47026	1.59
	implanted	0.29550	0.46906	1.59
MG-titanium ($Ti_{1.5}$)	initial	0.29529	0.46969	1.59
	implanted	0.29565	0.46888	1.59
MPC-titanium (Ti_{17})	initial	0.29537	0.46891	1.59
	implanted	0.29543	0.46892	1.59

remove oxygen and also in the precipitation of titanium oxides. As shown later, the formation of the oxides is detected at the dislocations in the volume of the titanium grains.

The supersaturation of the surface layers of titanium with oxygen in the conditions of ion implantation may also be caused by the existence of the surface titanium oxide film [71], which adsorbed on the surface the atomic concentration of this element tens of times greater. Theoretical studies of the processes of formation of phases of the oxide particles showed that the increase of the size of the particles with the increase of the depth of the implanted layer is determined by a decrease of the degree of supersaturation of the solid solution in the reaction zone [72]. The controlling factor is the relationship between the rate of diffusion transport of oxygen to this zone and the rate of formation (nucleation rate) of the oxide particles.

6.4. Effect of ion implantation on the phase composition of the surface layers of titanium alloys

The investigations carried out by transmission electron microscopy showed that in the initial condition all the structural states of the titanium targets consist of the grains of the α-Ti phase, characterised by the HCP crystal lattice and having the spatial group P63/mmc. In addition to the grains of α-titanium, the structure of the alloy contains small amounts (1–3 vol.%) of the grains of β-Ti (see the diagram in Fig. 6.1). The presence of the β-Ti phase in the commercial purity titanium of the VT1-0 grade, investigated in the present work (both coarse- and ultrafine-grained) was also reported previously in a number of investigations. These data have been generalised in [73].

The β-Ti phase has the BCC crystal lattice (spatial group Im3m) and has the form of plate-shaped precipitates. In the ultrafine-grained titanium ($Ti_{0.3}$) the β-Ti precipitates are distributed along the longitudinal boundaries of the α-Ti grains. Their size equals 50×200 nm. The volume fraction is ~1 vol.%. In MPC titanium (Ti17) these are parallel plates-shaped precipitates, situated inside the grains of α-Ti. The width of the individual plates is on average 50±10 nm, the length 1300±300 nm. The typical image of the α-Ti grain, containing the plate-shaped precipitates of β-Ti, is shown in Fig. 6.6. Analysis of the bright field (Fig. 6.6a) and dark field (Fig. 6.6b) images and the microdiffraction pattern, produced from this region (Fig. 6.6c, d) shows that, firstly, the precipitation of the β-Ti plates in relation to the lattice of α-Ti takes place in the $[110]_\alpha$ direction (this direction is indicated by the arrowhead in Fig. 6.6b). Secondly, the microdiffraction pattern shows two planes: the plane (001) of α-Ti and the plane (111) of β-Ti. This means that the axis of the [001] zone of α-Ti is parallel to the axis of the [111] β-titanium. Thirdly, the reflections [1$\bar{1}$0] α-Ti and [$\bar{1}$10] β-Ti on the microdiffraction pattern coincide. Thus, the crystal lattices of α-Ti and β-titanium are oriented so that the directions [1$\bar{1}$0] α-Ti and [$\bar{1}$10] β-Ti are parallel. All this indicates that the Jack orientation relationship is satisfied between the crystal lattices of α-Ti (HCP) and β-Ti (BCC). This is a typical relationship between the HCP and BCC crystal lattices in the α-Ti → β-Ti transformation [74]. According to the Jack relationship, the following relationship should be satisfied for the given case: $(001)_{\alpha\text{-Ti}}$ II $(111)_{\beta\text{-Ti}}$ and [1$\bar{1}$0]α-Ti II [$\bar{1}$10] β-Ti, as confirmed by transmission electron microscopy studies (Fig. 6.7).

Investigation of the phase composition of the implanted materials was carried out in two regions of the ion-doped layer: 1. Region I – the surface region, which depending on the grain size is found at a depth of up to 200 – 450 nm from the implanted surface were, according to the results obtained in Auger spectroscopy (Fig. 6.5) the content of the alloying elements is maximum; 2) the region II – the

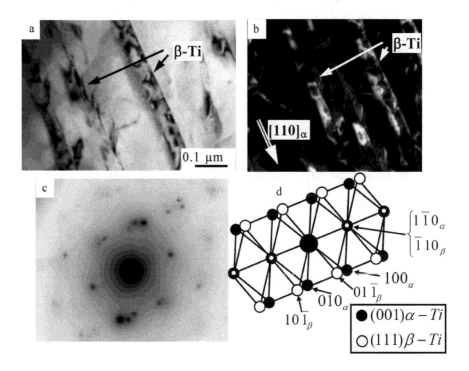

Fig. 6.6. Typical images of the fine structure of the grain of α-Ti of the second type, containing plate-shaped precipitates of β-Ti (indicated by the black arrows in (a, and by white arrows in (b). Initial condition of micro-polycrystalline titanium (Ti$_{17}$): a) bright field image; b) dark field image obtained in the [110] reflection; c) microdiffraction pattern; d) the indexed scheme.

central region of the ion-doped layer is situated, according to the Auger spectroscopy data (Fig. 6.5), at a depth of up to 650–850 nm.

According to the results of transmission electron microscopy studies, the α-Ti phase remains the main and dominant phase in all

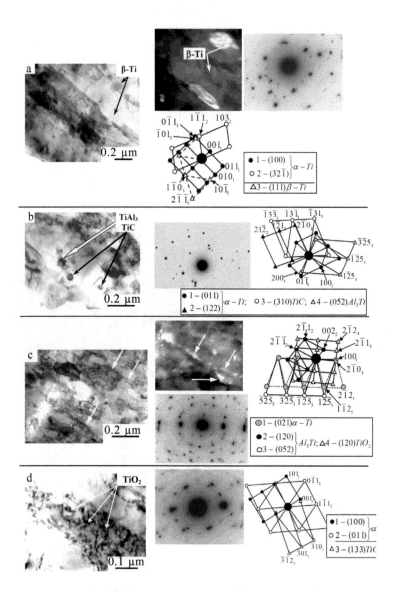

Fig. 6.7. Transmission electron microscope images of the secondary phases present in the region I of ultrafine-grained titanium ($Ti_{0.3}$); a) grains of β-Ti; b) the $TiAl_3$ phase and TiC particles; c) TiO_2 particles at the grain boundaries, and b) TiO_2 particles inside the α-Ti grains at the dislocations.

the implanted layers (up to 90–95% of the volume of the material). The α-Ti phase has the HCP lattice (the spatial group P63/mmc). In the specimens of ultrafine-grained titanium ($Ti_{0.3}$) the matrix grains of α-Ti have the anisotropic form determined, as mentioned previously, by the method of preparation of the specimen: multiple uniaxial pressing (abc-pressing) followed by multipass rolling in grooved rolls.

The experimental results show that the implantation of aluminium in the ultrafine-grained titanium ($Ti_{0.3}$) resulted in the formation of an entire set of phases characterised by different crystal lattices. The areas of localisation of the observed phases and their form in the thickness of the implanted layer are shown schematically in Fig. 6.2d, h. As indicated by the graph, the phase composition and the areas of localisation of the phases in the alloy change with increase of the distance from the irradiated surface (I → II).

In the region I (at a depth of 0–200 nm from the irradiated surface), in addition to the grains of α-Ti the structure of the alloy contains secondary phases. Firstly, it is the grains of the β-Ti phase with the BCC crystal lattice (spatial group Im3m). These grains are situated at the grain boundaries of α-Ti, have the anisotropic form and are small (70×250 nm). The volume fraction of these grains is small and equals ~1% in the volume of the material. Electron microscopic images of the grains of β-Ti are shown in Fig. 6.7a. The second phase, present in the material, is TiAl. This is the ordered phase with the DO_{22} superstructure, having the BCC crystal lattice with the spatial group 14/mmm. The $TiAl_3$ phase is represented by the circular nanograins distributed inside the α-Ti grains (Fig. 6.7b). The average grain size is 50 nm, volume fraction 2%.

The next phase is represented by the inclusions of the titanium carbide particles (TiC). This carbide has the FCC crystal lattice (spatial group Fm3m). The TiC particles are distributed inside the grains of α-Ti (Fig. 6.7b), their average size is 40 nm, volume fraction is ~1%. Another phase, present in the material, is represented by the inclusions of the titanium oxide TiO_2 (brookite), with the orthorhombic crystal lattice (spatial group Pbca). The TiO_2 particles are circular and distributed at the boundaries (Fig. 6.7c) and at the dislocations inside the grains of α-Ti (Fig. 6.7d). Their size is 10–20 nm, the total volume fraction in the material ~1%.

In region II (the central part of the ion-doped layer (250–450 nm from the irradiated surface)) the $Ti_{0.3}$ alloy also contains secondary phases, as in the region I. In particular, these phases are: β-Ti,

TiAl$_3$, TiC and TiO$_2$. Only their average size, form and areas of formation change (see the diagram in Fig. 6.3). The β-Ti grains are also distributed at the boundaries of the α-Ti grains and are characterise by the anisotropic form, but their size is smaller (60×130 mm). Their volume fraction also decreases (to 0.5%). The nanograins of the TiAl$_3$ ordered phase are no longer distributed inside the grains but they are found at the grain boundaries of α-Ti. Their mean size is greater (100 nm), and the volume fraction equals 1.5% of the total volume of the material. At the same time, the mean size of the TiC carbide particles decreases to 10 nm and the volume fraction to 0.5%. The TiC particles are distributed only at the dislocations inside the α-Ti grains. The areas of formation and the average size of the TiO$_2$ oxide particles remain the same as previously. The main distinguishing feature of the phase composition of the region II of the implanted layer is the presence of the Ti$_3$Al phase. This phase is an ordered phase with the D0$_{19}$ superstructure, having the HCP crystal lattice, its spatial group is P63/mmc. Ti$_3$Al is found in the material in the form of anisotropic nanograins at the boundaries of the α-Ti grains. The size of the grains is 25×70 nm, volume fraction 2.5%. The electron microscopic image of the Ti$_3$Al phase is shown in Fig. 6.8.

Thus, the implantation of Al in Ti$_{0.3}$ resulted in the formation of hardening phases such as β-Ti, TiAl$_3$, Ti$_3$Al, TiC and TiO$_2$, with the total content of these phases in the region I being ~5% of the volume of the material, and in the region II ~10%. This is in good agreement with the chemical analysis results, presented in Fig. 6.5.

The phase composition of the fine-grained titanium (Ti$_{1.5}$) was examined in detail. It should be mentioned that this specimen was produced from nanostructured titanium by successive annealing at 773 K. The secondary phases, formed in the implantation conditions, were TiO$_2$, Ti$_3$Al and TiAl$_3$ (see the diagram in Fig. 6.3). The area of formation of the phases depends on the distance in the implanted layer from the target surface. In the region I the secondary phases were titanium oxide TiO$_2$ (brookite) and the Ti$_3$Al ordered phase. The titanium oxide TiO$_2$ particles were situated at dislocations in the form of nanoparticles (average size of 20 nm) or in the form of plate-shaped precipitates (30×100 nm) inside the α-Ti grains (Fig. 6.9). The total volume fraction of the TiO$_2$ phase was ~2.5%. The formation of the Ti$_3$Al ordered phase in the form of anisotropic particles (average length 200 nm, average width 60 nm) was detected

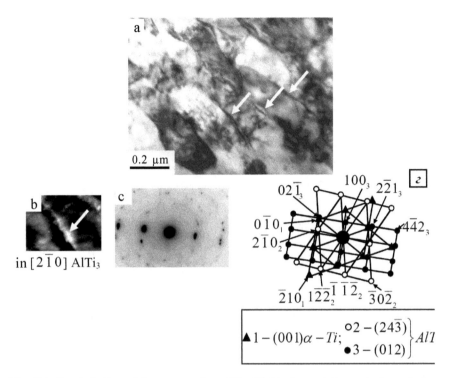

Fig. 6.8. Precipitation of the nanograins of the Ti$_3$Al phase at the grain boundaries of α-Ti. Transmission electron microscope images of the implanted layer of Ti$_{0.3}$ (the region II) a) bright field image, b) the dark field image, produced in the [4$\bar{4}$2] reflection of the AlTi$_3$ phase; c) microdiffraction pattern, and d) its indexed scheme (the planes: ▲ – α-Ti (001), o – Ti$_3$Al (243), ● – Ti$_3$Al (012) are present).

at the boundaries of the α-Ti grains (Fig. 6.10). The volume fraction of this phase was small, ~1%.

In the region II of the ion-doped layer of fine-grained titanium (Ti$_{1.5}$) as in the region I, there were particles of TiO$_2$ (brookite) titanium oxide and the Ti$_3$Al ordered phase. Only their average size, volume fraction and shape changed. In particular, the nanograins of the Ti$_3$Al phase were characterised by the isotropic form, with the size of 60 nm, smaller than in the region I, but their volume fraction was considerably greater (2%). There are distributed mostly at the joints of the large grains of α-Ti (Fig. 6.11). As for the TiO$_2$ oxide particles, the shape and areas of formation of these particles were the same as in region I (Fig. 6.12), only the average size and volume fractions were greater (~6%).

The main distinguishing feature of the phase composition of the region II of the implanted layer was the presence of the TiAl phase.

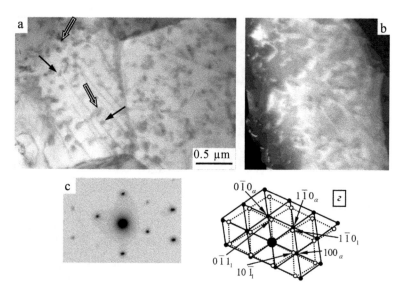

Fig. 6.9. Transmission electron microscope image of fine-grained titanium ($Ti_{1.5}$) after ion implantation, region I. The TiO_2 precipitates distributed inside the α-Ti grains, bright field image (a), dark field image (b) produced in the reflections [100] α-Ti + [$10\bar{1}$]TiO_2, microdiffraction pattern (c) and its indexed scheme (d), there are reflections belonging to the (001) α-Ti plane and (111) TiO_2 plane. In (a) the black arrows indicate the examples of spherical TiO_2 particles, composite arrows – plate-shaped particles.

The $TiAl_3$ phase is the ordered phase with the $D0_{22}$ superstructure, with the BCC crystal lattice. This phase was found in the material in the form of the nanograins of the anisotropic shape at the boundaries of the α-Ti grains (Fig. 6.11). The average size of the nanograins was 20×60 nm, the volume fraction in the material up to ~1%.

Thus, the implantation of aluminium in the fine-grained titanium ($Ti_{1.5}$) resulted in the formation of hardening phases, such as Ti_3Al, $TiAl_3$ and TiO_2, with the total amount of these phases in region I being 3% of the volume of the material, and in the region II 10%. This is in good agreement with the chemical analysis results presented in Fig. 6.5b.

According to the Auger electron spectroscopy results (Fig. 6.5a) the surface layer of the implanted MPC titanium (Ti_{17}) has the thickness of the modified layer of ~900–1000 nm with the concentration of the implanted impurity Al up to 70 at.%. The transmission electron microscopic studies of the implanted layer showed that the implantation of aluminium in MPC titanium resulted in the change of the phase composition in the surface layer. The

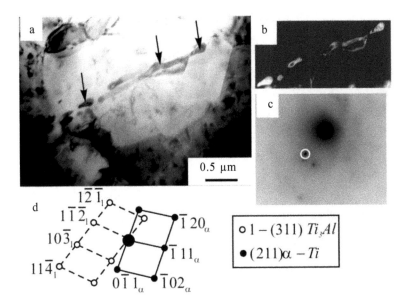

Fig. 6.10. Transmission electron microscope images of the ion-doped layer of fine-grained titanium ($Ti_{1.5}$) at a depth of up to 200 mm from the irradiated surface (the region I): bright field image (a), dark field image (b) produced in the [$01\bar{1}$] reflection of the Ti_3Al phase, indicated on the microdiffraction pattern (c) by the white circle, the diagram of the microdiffraction pattern (d). The precipitation of the Ti_3Al phase, distributed at the boundaries of the α-Ti boundaries, are indicated by the arrows (a).

results also showed that the change of the phase composition of the material takes place with increase of the distance from the irradiated surface and, namely, the set of the phases, the size of the phase particles, volume fractions area of formation change.

In region I (0–100) nm from the irradiated surface the structure contained grains of two types: type 1 – the grains of α-Ti with intrusions of particles of TiO_2 titanium oxide (Fig. 6.13), with the size of 10–50 nm, volume fraction in the material 0.3%; type 2 – the grains of α-Ti with the particles with the composition β-Ti and the Ti_3Al ordered phase. The crystal lattices of α-Ti (HCP) and β-Ti (BCC) are connected by the Jack orientation relationship, as in the initial material. The form of the particles of the Ti_3Al and β-Ti phases is the same, i.e., parallel particles of the plate-shaped form distributed inside the α-Ti grains. The width of the individual plates of the Ti_3Al phase on average was 50 ± 10 nm, length 1000 ± 100 nm. The average size of the individual particles of β-Ti in the region I was the same as in the initial material. The volume fraction of the

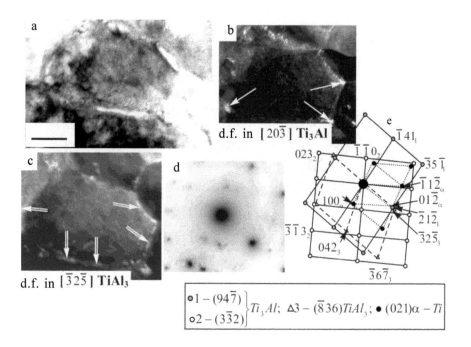

Fig. 6.11. Transmission electron microscope images of the ion-doped layer of fine-grained titanium (Ti$_{1.5}$) at a depth of 200–500 nm from the irradiated surface (the region II): bright field image (a), dark field image (d.f.) produced in the reflection of the Ti$_3$Al (b) and TiAl$_3$ (c) phases, the microdiffraction pattern (d) and its indexed scheme (e).

Ti$_3$Al phase in the region I was 1 vol.%, the fraction of the phase β-Ti was 2.5 vol.%. It should be mentioned that the fraction of the β-Ti phase in the initial material was 5 vol.%. This indicates that in the region I implantation resulted in the partial transformation of the ^Iβ-Ti-phase to the Ti$_3$Al phase. With increase of the distance from the surface into the implanted layer the amount of the Ti$_3$Al-phase increased and that of the ^Iβ-Ti-phase decreased, and at a depth of 300–400 nm (in the region II – the central region of the implanted layer) (Fig. 6.14a), i.e. in the region where the concentration of implanted aluminium alloy was maximum, the β-Ti phase was completely absent. The particles of the Ti$_3$Al-phase became larger with increase of the distance from the surface, and the shape of the particles remained plate-shaped. The particles of Ti$_3$Al were no longer present only inside the α-Ti grains but also at the boundaries of these grains. The average width of the plate in the region II was 80±20 nm, and the size of these particles was comparable with the

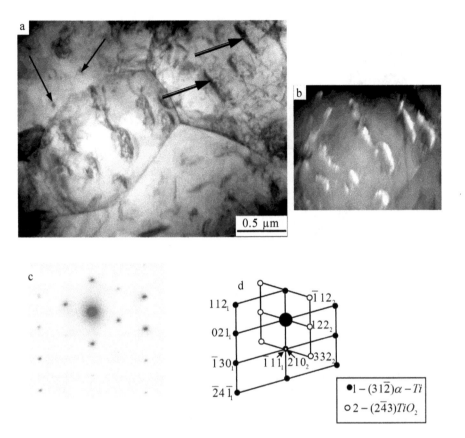

Fig. 6.12. Transmission electron microscope image of fine-grained titanium (Ti$_{1.5}$) after ion implantation, region II. TiO$_2$ precipitates distributed inside the grains of the α-Ti phase, bright field image (a), dark field image (b) produced in the [$\bar{1}$12] reflections of TiO$_2$, the microdiffraction pattern (c) and its indexed scheme (d), there are reflections belonging to the (31$\bar{2}$) plane of α-Ti and (2$\bar{4}$3) plane of TiO$_2$. In (a) the black arrows indicate examples of the circular particles of TiO$_2$, the double arrows – the plate-shaped particles.

size of the α-Ti grains, the volume fraction was maximum, equal to 5 vol.%. With a further increase of the distance from the surface into the implanted layer (with a decrease of the concentration of the implanted aluminium atoms) the amount of the Ti$_3$Al-phase decreases. At a distance of 900–1000 nm (the end of the implanted layer), i.e., in the area where the concentration of the implanted aluminium atoms reaches 3–5 at.%, the Ti$_3$Al-phase is completely absent, but the β-Ti phase is found.

In type 1 grains (grains of α-Ti with the intrusions of the titanium oxide particles TiO$_2$) with increase of the distance from the implanted

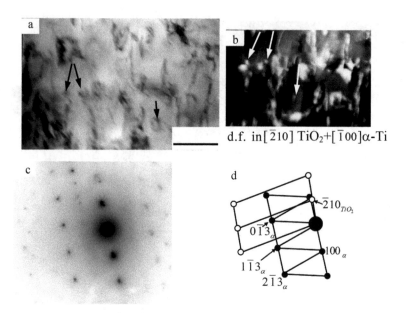

Fig. 6.13. Transmission electron microscope images of type 1 grains (grains of α-Ti with intrusions of the particles of titanium oxide TiO_2) of MPC-titanium (Ti_{17}) after ion implantation, region I. The precipitates of TiO_2 particles, distributed along the grains of α-Ti, bright field image (a), the dark field (d.f.) image (b) produced in the reflections [$\bar{1}00$] of α-Ti + [$\bar{2}10$] TiO_2, the microdiffraction pattern (c) and its indexed scheme (d), there are reflections belonging to the (031) plane of α-Ti and (243) plane of TiO_2 (examples of TiO_2 particles are indicated by the arrows).

surface into the thickness of the material the phase composition does not change – as previously, there are only particles of the TiO_2 oxide (circular particles). Changes are observed only in the shape and volume fraction of the particles (Fig. 6.14b): in the central part of the implanted layer the diameter of the particles is 35 nm, the volume fraction is maximum, 2.8 vol.%; at the end of the implanted layer the particle diameter is 40 nm, volume fraction 2 vol.%.

Thus, the phase composition of the ion-doped layers of titanium specimens with different grain sizes (0.3, 1.5 and 17 μm) was investigated and the regions of localisation of the aluminium phases in relation to the grains of the titanium matrix were determined (see the diagrams in Figs. 6.2–6.4). The base of the ion-modified layers are the grains of α-Ti. In the central part of the ion-implanted layer (in the region II) the formation of the Ti_3Al-phase was observed in all three specimens mostly in the form of plate-shaped precipitates. The experimental results show that the size of the produced Ti_3Al interlayers depends on the initial grains of the titanium matrix

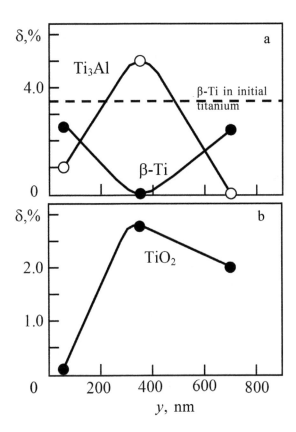

Fig. 6.14. Variation of the volume fraction of the secondary phases in the ion-doped layer of MPC titanium (Ti_{17}) in the thickness of the surface layers: a – β-Ti and Ti_3Al; b – TiO_2.

(Fig. 6.15a). As the grain size of the titanium target decreases, the thickness of the Ti_3Al interlayers at the grain boundaries also decreases (Fig. 6.15a). In the titanium targets with the large grains (17 μm) there is the ordered Ti_3Al phase localised in the form of interlayers not only at the boundaries of the α-Ti grains but also in the body of these grains with the size comparable with the matrix grains (Fig. 6.15a). An increase of the size of the matrix grains of α-Ti is accompanied by an increase of the longitudinal length of the particles of the Ti_3Al-phase (Fig. 6.15a). In the alloy with the small grain size (0.3 μm) the Ti_3Al-phase forms only at the grain boundaries. With an increase of the mean grain size of the matrix grains of α-Ti the amount of the phase, localised in the volume of the matrix, increases (Fig. 6.15c).

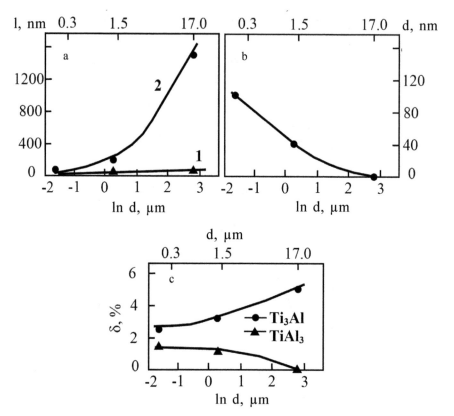

Fig. 6.15. Dependence of the dimensions (1 – longitudinal, 2 – transverse) of the particles of the second phases Ti$_3$Al (a) and TiAl$_3$ (b) and their volume fractions (c) on the average grain size of the titanium target (d).

The TiAl$_3$ phase forms in smaller quantities (no more than 1.5 vol.%) in comparison with the Ti$_3$Al phase. The particles of this phase are found only in titanium targets with the average size of the matrix grains of 0.3 and 1.5 μm, i.e., in ultrafine-grained and fine-grained titanium (see the diagrams in Figs. 6.2 and 6.3). In most cases, the particles of the TiAl$_3$ phase are equiaxed and localised at the grain boundaries of the titanium matrix. The dimensions and the volume fraction of the TiAl$_3$ phase decreases with increase of the size of the matrix grains of α-titanium (Figs. 6.15b, c). A significant role in the localisation of the TiAl$_3$ phase is played by the defective structure of the material. The addition of aluminium results in the stabilisation of the titanium matrix and supports the formation of the TiAl$_3$ ordered phase at the boundaries of α-titanium grains.

According to the Ti–Al equilibrium diagram [75] the concentration of the implanted impurity required for the formation of the TiAl$_3$ phase is ~70%. At the same time, according to the Auger spectroscopic data, the aluminium concentration does not exceed 60 at.% (Fig. 6.5a). The refining of the titanium grains initiates the increase of the number of channels for the formation of new phases as a result of easier diffusion of the alloying elements. The substitutional element – aluminium – travels during ion implantation along the grain boundaries or through the solid solution. The high-energy boundaries of the titanium grains are characterised by the higher aluminium concentration leading to the formation of the TiAl phase. In addition, ion implantation took place at 623 K and this increased the diffusion mobility of the aluminium atoms. Previously, the results obtained by the authors of the present book [16] showed that at a high aluminium concentration (up to 75 at.%) the TiAl$_3$ phase did not form in the mesopolycrystalline titanium. Correspondingly, the formation of the TiAl$_3$ phase is strongly affected by defectiveness, in particular, the presence of a high concentration of triple junctions at the grain boundaries which act as sinks for impurities. Additional stabilisation of the structure of the material with the parameters of the grains in the microregion takes place by pinning of the grain boundaries and their junctions by intermetallic phases and this supports extensive hardening of the titanium materials. It should be noted that in region I of the ion-doped layers of all alloys the size and volume fractions of the resultant aluminium phases are smaller than in the region II, but the tendency with increasing grain size is the same.

Previously, it was shown that the implantation of aluminium in titanium resulted in the formation of titanium carbides and oxides. It was also established that the size and volume fraction of the particles depend on the grain size of the titanium target. According to the Auger spectroscopic data, the carbon concentration of the alloy with the smallest grain size (0.3 µm) in the entire ion-implanted layer was 5–8 at.% (Fig. 6.5 a), and in the alloys with a grain size of 1.5 and 17 µm it was only 2–3 at.% (Fig. 6.5b, c). Therefore, the TiC carbide forms only in the ultrafine-grained alloy. The TiO$_2$ oxides formed in all alloys and in the entire thickness of the ion-doped layers. This is also in good agreement with the Auger spectroscopic data (Fig. 6.5) – the oxygen concentration is almost an order of magnitude higher than that of carbon. The maximum amount of the oxides was found in the fine-grained titanium, the minimum in the MPC titanium

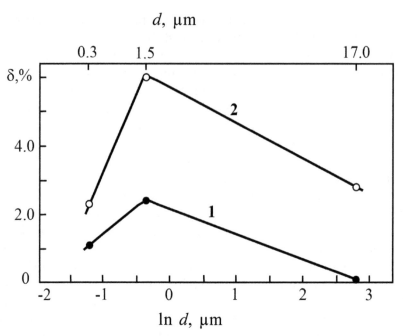

Fig. 6.16. Dependence of the volume fraction of the TiO_2 phase (δ) on the average grain size of the titanium target (e): 1 – region I; 2 – region II of the ion-doped layer.

(Fig. 6.16). In region II the number of oxides in all the alloys was greater in comparison with the region I. However, the tendency of the dependence of the volume fraction of the TiO_2 phase on the grain size of the target in both region was that same (Fig. 6.16).

6.5. Modification of the physical–mechanical properties of titanium alloys by the ion implantation conditions

The typical characteristics of the quantitative parameters of the mechanical properties of the metallic materials are in particular the yield strength and strain hardening. In addition to this, one of the quantitative parameters of the mechanical properties of the metallic materials is also microhardness. It should be mentioned that the microhardness is the resistance of the material to elastic and plastic deformation in penetration or scratching with a rigid indentor and characterises generally the mechanical strength of the crystal lattice in the conditions of the complex and inhomogeneous volume stress state. The quantitative investigations of the parameters of the structure of the investigated alloys prior to and after implantation can be used to investigate the process of redistribution of the aluminium atoms, changes of the phase composition, the redistribution of the density of dislocations and internal stresses, and estimate the yield strength of the implanted alloys. The results can also be used to explain quite efficiently the behaviour of microhardness.

At present time, it has been established that the strength of any metallic material is determined by many factors [77], the main of which are: 1) the presence of the grain boundaries and other structural formations; 2) high dislocation density formed in any effect on the material; 3) the presence in the material of carbide, oxide particles and the secondary phases, etc. The role of each factor in its specific case will differ, and the contribution of the individual hardening mechanisms to the total hardening of the material will also differ.

The results of a large number of theoretical and experimental studies have been used to derive a large number of relationships for determining the quantitative relationship of the parameters of the fine structure of the material which its mechanical properties [76–78]. In this work, according to the well-known assumptions [76–78], the yield strength was calculated using equation (6.1) which sums up quadratically the contributions of hardening of the 'forest' dislocations and the internal fields, with the other contributions added up:

$$\sigma = \Delta\sigma_n + \Delta\sigma_m + \Delta\sigma_g + \Delta\sigma_{or} + \sqrt{\left(\Delta\sigma_{long} + \Delta\sigma_f^2\right)}. \qquad (6.1)$$

Equation (6.1) describes almost all contributions to the deformation resistance. In this equation $\Delta\sigma_n$ is the friction stress of the dislocations in the crystal lattice of α-Ti; $\Delta\sigma_s$ is the hardening of the solid solution based on α-Ti by the atoms of the alloying elements (Al, C, O); $\Delta\sigma_f$ is the hardening by the 'forest' dislocations which intersect the colliding dislocation; $\Delta\sigma_{long}$ is the hardening by the long-range stress fields; $\Delta\sigma_{or}$ is the hardening of the material by non-coherent particles when they are bypassed by the dislocations by the Orowan mechanism; $\Delta\sigma_g$ is the hardening by the grain boundaries.

The role of each factor will be examined in greater detail for the initial states of the investigated alloys ($Ti_{0.3}$, $Ti_{1.5}$, Ti_{17}). As mentioned previously, in the initial condition regardless of the average grain size all the alloys are almost completely single-phase alloys – α-Ti. Therefore, the contribution to the hardening of the material by the presence of the third factor and, in particular, the presence of secondary phases, in all the investigated alloys will be the same and minimum. It should be noted that the presence of the β-Ti phase in the $Ti_{0.3}$ and Ti_{17} alloys does not provide any significant contribution to hardening because the volume fraction of this phase does not exceed 1–2% of the volume of the material. The contribution to the hardening of the material of the presence of the second factor, namely the dislocation density, is also small (40–60 MPa) and almost the same, because the scalar dislocation density in all alloys was similar and equalled $(0.1–0.2) \cdot 10^{14}$ m^{-2}. There is no hardening by the long-range stress fields. The components $\Delta\sigma_n$ and $\Delta\sigma_m$ also provide only a minimum contribution to the general hardening because, according to the Auger spectroscopic data obtained by the authors of the present book [57] and the x-ray diffraction analysis data [79], the solid solution of the initial alloys did not contain any carbon and oxygen. The most important factor which determines the hardening of the alloys investigated in the initial state is the first factor, namely the presence of the grain boundaries or, in other words, the density of the grain boundaries. As shown by the experiments, the yield strength of the material is associated with the grain size d (grain boundary hardening). The strength increases as a result of the grain boundaries of α-Ti which represent barriers to the propagation of yielding. It is described by the well-known Hall–Petch relationship [80]:

$$\Delta\sigma_g = k \times d^{-0.5}, \tag{6.2}$$

where k is the proportionality coefficient which depends on the purity of the material, the extent of deformation of the material, the type and structure of the grain boundaries, the hardening of the boundaries by the second phase particles and on the grain size [24, 77] (the average value of coefficient k for α-Ti is given in Table 6.2 [24]). The grain boundary hardening in the initial alloys is determined mainly by the grain size. This means that the hardening should be higher in the alloy in which the grain boundary density is higher (the grain size is smaller), i.e., in the ultrafine-grained titanium alloy ($Ti_{0.3}$). Table 6.2 shows the results of grain boundary hardening for different grain sizes investigated in the study. It may be seen that the highest value of $\Delta\sigma_g$ is obtained for ultrafine-grained titanium ($Ti_{0.3}$), in fine-grained titanium ($Ti_{1.5}$) it is approximately half the contribution, and in MPC titanium (Ti_{17}) it is more than four times smaller.

Table 6.2. The Hall–Petch relationship and the magnitude of grain boundary hardening in the investigated alloys prior to and after implantation

Specimen, average grain size	Hall-Petch coefficient, k, MPa $m^{1/2}$	Grain boundary hardening $\Delta\sigma_g$, MPa		
		Prior to implantation	After implantation	
			Region I	Region II
$Ti_{0.3}$ (0.3 μm)	0.25	429	559	573
$Ti_{1.5}$ (1.5 μm)	0.30	263	316	263
Ti_{17} (17 μm)	0.35	85	90	85

The effect of the factors for the implanted states of the titanium alloys is different. Since implantation slightly changes the grain size in the ion-doped alloys, the magnitude of grain boundary hardening changes, but not very significantly (Table 6.2). An important role in the hardening of the implanted alloys is played by the forest dislocations, both for the non-polarised (for the non-charged dislocation ensemble) and polarised alloys (or the charged dislocation ensemble) [78]. The non-charged dislocation ensemble is the ensemble without excess dislocations, when $\rho_+ = \rho_-$ (here ρ_+ is the density of positively charged dislocations, ρ_- is the density of negatively charged dislocations), and consequently the excess dislocation density $\rho_\pm = \rho_+ - \rho_- = 0$. In this case, as already mentioned previously, the non-charged dislocation ensemble generates a shear stress (stress fields, produced by the dislocation structure) determined using the equation [76]:

$$\Delta\sigma_f = m\alpha Gb\sqrt{\rho}, \qquad (6.3)$$

where m is the orientation multiplier; α is a dimensionless coefficient which changes in the range 0.05–1.0 depending on the type of dislocation ensemble [78]; G is the shear modulus of the material of the matrix; b is the Burgers vector; ρ is the mean value of the scalar dislocation density. In the charged dislocation ensemble when the excess dislocation density is:

$$\rho_\pm = \rho_+ - \rho_- \neq 0,$$

there are moment (long-range) stresses of magnitude determined by electron microscopy on the basis of the bending extinction contours observed in the structure of the material [76, 77]. It is calculated from the equation

$$\Delta\sigma_{long} = m\alpha_S Gb\sqrt{\rho_\pm}, \qquad\qquad (6.4)$$

where $\alpha_S = 1$ is the Strunin coefficient [81]. As mentioned previously, implantation resulted in the entire ion-doped layer in the buildup of high (in comparison with the initial state) scalar dislocation density (Table 6.3). This dislocation structure in turn formed high shear stresses whose magnitude is also shown in Table 6.3. As indicated by the table, the highest shear stresses formed in the ultrafine-grained titanium ($Ti_{0.3}$), the lowest stresses in fine-grained titanium ($Ti_{1.5}$). Evidently, this is associated with phase formation: in the ultrafine-grained titanium ($Ti_{0.3}$) there is the largest set of the secondary phases, in fine-grained titanium ($Ti_{1.5}$) – the smallest set.

Table 6.3. Dislocation structure parameters after implantation

Specimen, average grain size	No. of region of ion-doped layer	Non-charged dislocation ensemble		Charged dislocation ensemble	
		Scalar dislocation density $\rho \cdot 10^{-14}$, m^2	Shear stress, $\Delta\sigma_f$ MPa	Excess dislocation density $\rho_\pm \cdot 10^{-14}$, m^2	Moment (long-range acting) stresses, $\Delta\sigma_{long}$, MPa
$Ti_{0.3}$ (0.3 μm)	I	8.5	410	0	0
	II	7.20	375	0	0
$Ti_{1.5}$ (1.5 μm)	I	0.8	115	0.5	230
	II	2.3	210	2.3	650
Ti_{17} (17 μm)	I	2.6	230	2.2	210
	II	6.3	360	4.4	300

Implantation resulted in the polarisation of the dislocation structure (the electron microscopic images do not show the bending extinction contour's) and, correspondingly, in the formation of moment (long-range) stresses, by different mechanisms in different alloys (Table 6.3). For example, in ultrafine-grained titanium ($Ti_{0.3}$) the value of ρ is the highest (Table 6.3), but there is no polarisation of the dislocation structure (the structure does not contain bending extinction contours). In MPC (Ti_{17}) the value ρ is not so high but, nevertheless, the dislocation structure is almost completely polarised, since ρ_+ is close to ρ (Table 6.3).

Previously, it was mentioned that implantation resulted in the formation of a large set of phases. The results show that the set of the phases depends on the average grain size of the alloy. It is well-known [82] that the dispersion hardening of the alloy depends on the number of particles, the size of the particles, the type of distribution and the distance between the particles, and also the degree of mismatch of the crystal lattices of the matrix and the precipitation. The dispersion hardening mechanisms have been developed for non-coherent particles when the dislocations bypass the precipitates, and for the coherent particles when the dislocations may intersect the particles. In the alloys investigated in this work, the non-coherent particles were particles of TiC and TiO_2, the coherent particles – the particles of the phases β-Ti, Ti_3Al and TiAl. In addition to this, the dispersion hardening mechanisms are sub-divided into main mechanisms (the particles are distributed inside the grains) and indirect mechanisms (particles are situated at the boundaries and junctions of the grains). The main mechanisms increase the yield strength of the alloy, the indirect mechanisms influence the strength and hardness. In all these cases, the hardening can be calculated using the simplified Orowan equation [83]:

$$\Delta\sigma_{or} = 2\lambda Gb / r, \qquad (6.5)$$

where $\lambda = 0.5$ is a coefficient which takes into account all these moments, r is the distance between the particles. The results of calculations of $\Delta\sigma_{or}$ after implantation of aluminium in titanium are presented in Table 6.4 (the – indicates the absence of the phase in the alloy). The table shows that the main contribution to hardening in all alloys is provided by the TiO_2 oxide, and in the ultrafine-grained titanium ($Ti_{0.3}$) the contribution is maximum, and in the MPC titanium (Ti_{17}) it is the smallest.

The contribution to general hardening of the components $\Delta\sigma_n$ and $\Delta\sigma_m$ is minimum, as in the initial condition, because according to the Auger spectroscopy data and x-ray diffraction analysis results [57, 58] the concentration of aluminium, carbon and oxygen in the solid solution after implantation is very small.

Thus, the calculations carried out using equation (6.1) show (Fig. 6.17) that, firstly, both prior to and after implantation the increase of the grain size decreases the magnitude of hardening of the alloy (Fig. 6.17a). Secondly, the implantation of aluminium in all the alloys investigated in this work resulted in significant hardening of the

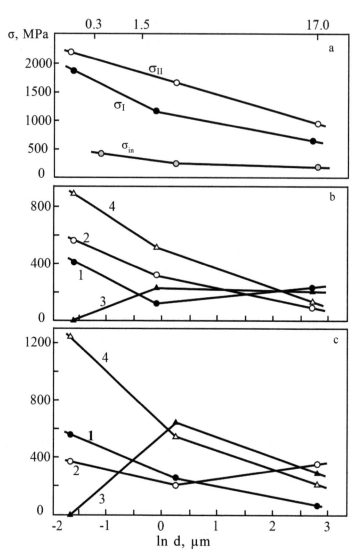

Fig. 6.17. Dependence of the yield strength σ and contributions of the individual hardening mechanisms on the grain size of α-Ti in the thickness of the ion-doped layer of titanium: a – σ_{in} (the yield strength of the initial material); b – in the region I; c – in the region II of the ion-doped layer (1 – $\Delta\sigma_g$, 2 – $\delta\sigma_p$, 3 – $\Delta\sigma_{long}$, 4 – $\Delta\sigma_{or}$).

alloys. In particular, the hardening in the ultrafine-grained titanium (Ti$_{0.3}$) and fine-grained titanium (Ti$_{1.5}$) increased almost 3 times, and in MPC titanium (Ti$_{17}$) by more than six times. The contributions of the individual components, determined using the equations (6.2)–(6.55) to the total hardening of the alloys are presented in Fig. 6.17b,

c. As indicated by the graphs, the magnitude of contribution of the individual hardening mechanisms to the overall hardening of the alloys differs. In addition, the contribution of each mechanism in each specific case is different.

Table 6.4. Dispersion hardening of implanted alloys by different phases

Specimen, average grain size	No. of region of ion-doped layer	$\Delta\sigma_{or}$, MPa				
		Ti_3Al	$TiAl_3$	β-Ti	TiC	TiO_2
$Ti_{0.3}$ (0.3 μm)	I	—	102	20	85	685
	II	160	30	20	405	635
$Ti_{1.5}$ (1.5 μm)	I	25	—	—	—	500
	II	100	140	—	—	310
Ti_{17} (17 μm)	I	15	—	20	—	100
	II	15	—	—	—	205

As indicated previously, microhardness was measured in the investigated alloys prior to and after implantation. Since the microhardness is the deformation resistance of the material, the results obtained previously for the hardening of the alloys were used for interpreting the behaviour of microhardness.

The experimental results show that a decrease of the grain size, i.e., increase of the density of the boundaries in the alloys, results in an increase of the microhardness of the initial alloys (Fig.s 6.18 and 6.19). It may be seen clearly that in ultrafine grained titanium ($Ti_{0.3}$) characterised by the highest density of the grain boundaries, the microhardness is the highest. In mesopolycrystalline titanium (Ti_{17}) with the largest grain size, i.e. the lowest grain boundary density, the microhardness is the lowest. The results are in good agreement with the results obtained for the hardening of these alloys. Therefore, at the level of the grain size, a significant role is played by the contribution of the high concentration of the grain boundaries to microhardness. In this case, the defective structure of the grain boundaries is the most important factor.

Ion implantation is followed by a significant increase of microhardness (Figs. 6.20 and 6.21), and the largest microhardness increase was observed in MPC titanium (Ti_{17}) (Fig. 6.21, curve 3). This is also in agreement with the previously described results.

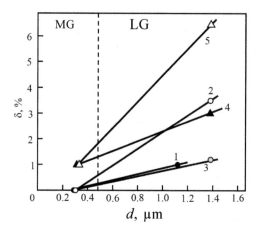

Fig. 6.18. Dependence of the volume fraction of the secondary phases Ti$_3$Al (1, 2), TiAl$_3$ (3) and TiO$_2$ (4, 5) at the grain size of α-Ti in the thickness of the ion-doped layer of titanium: 1, 4) to 200 mm from the irradiated surface (the region I); 2, 3, 5) at a depth of 200–500 nm from the irradiated surface (the region II).

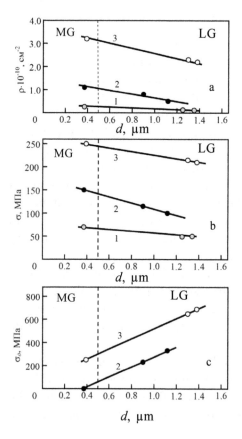

Fig. 6.19. Dependence of the scalar dislocation density (a), shear stress (b) and local (moment) stresses (c) and the average grain size α-Ti prior to and after ion implantation: 1 – the initial non-implanted titanium, 2 – the characteristics of the ion-doped layer in the region I, 3 – the characteristics in the region II of the surface layer.

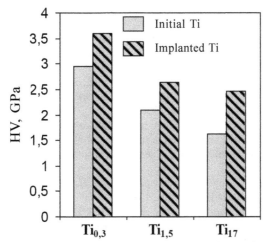

Fig. 6.20. Microhardness of the titanium specimens prior to and after ion implantation.

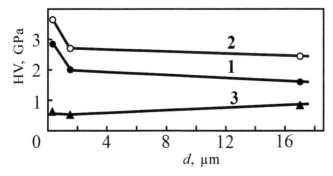

Fig. 6.21. Effect of grain size of microhardness (1 – prior to implantation; 2 – after implantation) and the increase of microhardness (3) in implantation of VT1-0 alloy with aluminium ions.

Therefore, it may be claimed that the increase of the microhardness of the implanted layers is associated with hardening as a result of the formation of nanocrystalline aluminide phases in the volume of the matrix grains. Additional stabilisation of the structure of the material with the parameters of the grains in the microregion takes place by the pinning of the grain boundaries and their junctions by intermetallic phases resulting in significant hardening.

References

1. Guseva M.I., *Poverkhnost. Fizika, khimiya, mekhanika*, 1982, No. 4, 27–50.
2. Ion implantation, translated from English. ed. O.P. Elyutin, Moscow, Metallurgiya, 1985.
3. Brown G., *Nucl. Instr. Methods*, 1989, V. B37–38, 68–70
4. Belyi A.V., et al., Surface hardening treatment with the use of concentrated energy fluxes, Minsk, Nauka i tekhnika, 1990.
5. Komarov F.F., Ion implantation into metals, Moscow, Metallurgiya, 1990.
6. Komarov F.F., Novikov A.P., Ion beam mixing of of irradiated metals. Itogi nauki i tekhniki. Seriya Puchki zaryazh. chastits i tverdoe telo, Moscow, VINITI, 1999, V. 7, 54–81.
7. Belyi A.V., et al., Ion beam processing of metals, alloys and ceramic materials, Minsk, Physico-Technical Institute, 1998.
8. Tsiganov I., et al., *Thin Solid Films*, 2000, V. 376, 188–197.
9. Gribkov V.A., et al., Prospective radiation beam materials processing technology, Moscow, Kruglyi god, 2001.
10. Komarov F.F., Physical processes of ion implantation into solids, Minsk, Tekhnoprint, 2001.
11. Bozhko I.A., Laws of formation of ultrafine intermetallic phases in the surface layers of nickel and titanium at high-intensity ion implantation, Dissertation, Tomsk, 2008.
12. Kurzina I.A., Gradient surface layers based on nanoscale metal particles: synthesis, structure and properties, Dissertation, Barnaul, 2011.
13. Kozlov E.V., et al., *Vestnik TGASU*, 2000, No. 2, 87–91.
14. Kurzina I.A., et al., *Izv. Tomsk. Politekh. Univ.*, 2004, No. 2, 30–35.
15. Sharkeev Yu.P., et al., *Izv. VUZ Fizika*, 2004, No. 9, 44–52.
16. Kurzina I.A., et al., *Metallofizika i noveishie tekhnologii*, 2004, No. 12, 1645–1660.
17. Kurzina I.A., et al., *Perspektivnye Materialy*, 2005, No. 1, 13–21.
18. Kurzina I.A., et al., *Poverkhnost'. Rentgenovskie, sinkhtronnye i neitronnye issled.*, 2005, No. 7, 72–78.
19. Kurzina I.A., et al., *Izv. RAN, Ser. Fiz.*, 2005, V. 69, No. 7, 1002–1006.
20. Kurzina I.A., et al., *Khimiya i fizika stekla*, 2005, V. 31, No. 4, 452–458.
21. Kurzina I.A., et al., *Izv. RAN, Ser. Fiz.*, 2006, V. 70, No. 4, 591–592.
22. Kozlov E.V., et al., *ibid*, 2007, V. 71, No. 2, 198–201.
23. Kurzina I.A., et al., *ibid*, 2008, V. 72, No. 2, 1208–1211.
24. Kurzina I.A., et al., Nanocrystalline intermetallic and nitride structures formed in ion-plasma exposure, Tomsk, NTL, 2008.
25. Sharkeev Yu.P., et al., in: Structural and phase states of promising metallic materials (ed. V.E. Gromov), Novokuznetsk, Publishing House of the NPC, 2009, 565–579.
26. Kurzina I.A., et al., *Fundamen. Probl. Sovremen. Materialovedeniya*, 2009, No. 2, 63–69.
27. Kurzina I.A., et al., *Materialovedenie*, 2010, No. 5, 48–54.
28. Sinel'nikova V.A., et al., Aluminides, Kiev, Naukova Dumka, 1965.
29. Andrievskii R.A., Ragulya A.V., Nanostructured materials, textbook, Moscow, Akademiya, 2005.
30. Greenberg B.A., Ivanov M.A., Intermetallic compounds Ni_3Al and TiAl: Microstructure, deformation behavior, Ekaterinburg, Izd. UroRAN, 2002.
31. Kurzina I.A., et al., in: Structure and properties of advanced materials, Tomsk: Publishing house of the NTL, 2007, 159–195.

32. Gunter V.E., et al., Medical materials and implants with shape memory, Tomsk, Publishing house of Tomsk State University, 1998.
33. Valiev R.Z., et al., *Konstruktsii iz Kompozit. Mater.*, 2004, No. 4, 64–66.
34. Kolobov Yu.R., et al., in: Physical mesomechanics and computer design of materials (ed. V.E. Panin), in 2 vols., Nauka, Novosibirsk, 1995. Vol. 2, 214–239.
35. Gritsenko B.P., et al., *Fiz. mezomekhanika*, 2004, V. 7 (spec. issue), Part 1, 415–418.
36. Kozlov E.V., et al.4, *ibid*, 2004, V. 7, No. 4, 93–111.
37. Kozlov E.V., et al., *ibid*, V. 9. No. 3, 81–92.
38. Koneva, N.A., et al., *Izv. RAN, Ser. Fiz.*, 2006, V. 70. No. 4, 582–585.
39. Kozlov E.V., et al., in: Nanomaterials by severe plastic deformation, Weinheim: Wiley-VCH Verlag GmbH and Co, KGaA, 2004, 263–270.
40. Kozlov E.V., et al.,*Fiz. mezomekhanika*, 2011, V. 14, No. 3, 95–110.
41. Kozlov E.V., et al., Pis'ma o materialakh, 2011, V. 4, 113–117.
42. Konev N.A., et al., *Perspekt. Materialy*, 2011, No. 12, 238–242.
43. Kozlov E.V., et al., Proc. XIII Materials International Meeting 'Radiation Physics of Solids', Sevastopol', 2003, 364–368.
44. Kurzina I.A., et al., in: 8th Korean–Russian International Symposium on Science & Technology KORUS 2004, Tomsk, 2004, 127–131.
45. Bozhko I.A., et al., in: XIV International Meeting 'Radiation Physics of Solids', Sevastopol', 2004, 357–361.
46. Kurzina I.A., et al., *ibid*, 362–366.
47. Kurzina I.A., et al., in: Conference on Modification of Materials with Particle Beams and Plasma Flows, Tomsk, 2004, 221–224.
48. Kurzina I.A., et al., in: 7th International symposium 'Phase Transformations in Solid Solutions and Alloys'. Sochi 2004, Rostov n/D, Publishing House of the RGPU, 2004, 178–181.
49. Kurzina I.A., et al., in: 8th International Symposium 'Phase Transformations in Solid Solutions and Alloys', Sochi 2005, Rostov n/D: Publishing House of the RGPU, 2005, Part 1, 212–214.
50. Knyazeva A.G., et al., in: VI International Conference 'Interaction of radiation with solids', Minsk, 2005, 39–41.
51. Kurzina I.A., et al., in: 8th Conference on Modification of Materials with Particle Beams and Plasma Flows, Tomsk, 2006, 211–214.
52. Sharkeev Yu.P., et al., in: Proceedings of the Symposium 'Ordering in metals and alloys', Rostov-on-Don, Rostov, 2006, 230–232.
53. Kurzina I.A., et al., in: Proceedings of the Russian school-conference of young scientists 'Biocompatible Nanostructured materials and coatings for medical purposes', Belgorod, 2006, 100–105.
54. Sharkeev Yu.P., et al. in: Proceeding of the Second Asian Symposium on Adv. Mater. Chemistry of Functional Materials, 2009, Shanghai, Fudan University, 51–53.
55. Kurzina I.A., in: Proceedings of the 14th International Symposium 'Ordering of minerals and alloys', 2011, Rostov-on-Don, Publishing house SKNTS VSH SFU APSN, 2011, 201–204.
56. Kurzina I.A., et al. in: Proceedings of the 15th International Symposium 'Ordering of minerals and alloys', 2011, Rostov-on-Don, Publishing house SKNTS VSH SFU APSN, 2012, 152–155.
57. Kurzina I.A., et al., *Izv. VUZ, Fizika*, 2011, V. 54, No. 11, 112–119.
58. Kurzina I.A., et al., *ibid*, 2012, V.76, No. 1, 74–78.
59. Kurzina I.A., et al., *Bulletin of the Russian Academy of Sciences. Physics.* 2012. V. 76, No. 1, 64–68.

60. Kurzina I.A., et al., *Izv. VUZ, Fizika*, 2012, V. 76, No. 11, 1384–1392.
61. Kurzina I.A., et al., *Fundament. Problems Sovremenn. Materialoved.*, 2012, V. 9, No. 4, 495–502.
62. Kurzina I.A., et al., *Bulletin of the Russian Academy of Sciences, Physics*, 2012. V. 76, No. 11, 1238–1245.
63. Kurzina I.A., et al., *Izv. VUZ, Fizika*, 2012, No. 5, 185–191.
64. Kurzina I.A., et al., *Fundament. Problems Sovremenn. Materialoved.*, 2012, V. 10, No. 1, 35–41.
65. Kozlov E.V., et al., *Vestnik Tambovsk. Univ.*, 2000, V. 8, No. 4, 509–514.
66. Kozlov E.V., et al., in: Control. Diagnostics. Service life (ed. V.Yu. Blumenstein and A.A. Krechetov), Kemerovo, 2007, 135–142.
67. Kozlov E.V., et al., *Voprosy Materialoved.*, 2008, No. 2 (54), 51–59.
68. Gushenets V.I., et al., *Review of Scientific Instruments*, 2006, V. 77, No. 6, Art. 063301.
69. Savkin K.P., et al., in: Proc. 9th Int. Conf. on Modification of Materials with Particle Beams and Plasma Flows, 2008, 68–71.
70. Synthesis and properties of nanocrystalline and substructural materials (ed. A.D. Karataev), Tomsk, TSU, 2007.
71. Bai A.S., et al., Oxidation of titanium and its alloys, Moscow, Metallurgiya, 1970.
72. Mejering J.E., *Adv. in Mater. Res.*, 1971, No. 5, 1–81.
73. Kashin O.A., Deformation behavior in microplastic deformation of Ti–Al–V alloy and titanium alloy with the ultrafine-grained structure for different types of thermal-power exposure, Dissertation, Tomsk, 2007.
74. Utevskii L.M., Diffraction electron microscopy, Moscow, Metallurgiya, 1973.
75. Diagrams of binary metallic systems (ed. N.P. Lyakishev), Moscow, Mashinostroenie, Vol. 1, 1996.
76. Koneva N.A., Kozlov E.V., *Izv. VUZ, Fizika*, 1991, No. 3, 56–70.
77. Kozlov E.V., Koneva N.A., *ibid*, 2002, No. 5. 52–71.
78. Koneva N.A., Kozlov E.V., in: Advanced materials (textbook, ed., D.L. Merson), Moscow Institute of Steel and Alloy, 2006, 267–320.
79. Kurzina I.A., et al., in: Proceedings of the 15th International Symposium 'Order, disorder and properties of oxides', Rostov-on-Don, 2012, Publishing house SKNTs VSh YuFU APSN, 2012, 180–182.
80. McLean D., Mechanical properties of metals [Russian translation], Moscow, Metallurgiya, 1965.
81. Strunin B.N., *Fiz. Tverd. Tela*, 1967, V. 9, No. 21, 805–812.
82. Honeycombe R.W., Plastic deformation of metals [Russian translation], Moscow, Mir, 1972.
83. Goldstein M., Farber V.M., Precipitation hardening of steel, Moscow, Metallurgiya, 1979.

7

Grain boundary engineering and superhigh strength of nanocrystals

In this chapter, attention is given to the new concepts of the strengthening of nanocrystals based on the competition of different plastic deformation mechanisms: dislocation slip and grain boundary gliding. It is well-known that the value of the theoretical (limiting) strength of the solid can be determined in fractions of the shear modulus [1]:

$$\sigma_{th} = \frac{G}{\alpha} = \frac{E}{2\alpha(1+\nu)}, \qquad (7.1)$$

where G and E are respectively the values of the shear modulus and normal elasticity modulus, ν is the Poisson coefficient, α is a numerical coefficient which is in the range $5 < \alpha < 30$, depending on the type of stress state and the calculation method [1]. The value σ_{th} is the physical ultimate strength which in principle can be obtained and is used as an orientation point in the development of actual high-strength materials.

On the other hand, the most important element of the structure of nanocrystalline materials, which determines their strength properties, are the grain boundaries (GB) [2]. The dependence of the yield stress σ_y on the average grain size in the single-phase polycrystalline aggregates is governed, as is well-known, by the Hall–Petch relation (H–P) [1]:

$$\sigma_y = \sigma_0 + K_y D^{-1/2} \tag{7.2}$$

where σ_0 is the yield stress of plastic flow in the body of the grain with the average size D, and K_y is the proportionality coefficient, characterising the 'transparency' of the GBs for plastic deformation.

Since the dependence (7.2) reflects the dislocation nature of the plastic flow of polycrystalline materials, it should be expected that it is universal. Consequently sooner or later, any material having the form of a polycrystalline ensemble of the grains may reach the theoretical strength σ_{th} at the point A with the decrease of D in the nanometre range (Fig. 7.1)

The equations (7.1) and (7.2) can be used to determine easily the average grain size of the polycrystalline ensemble D_A at which the strength of the polycrystal may correspond to the theoretical limit on the condition of the exact fulfilment of the H–P relation:

$$D_A = \left(\frac{K_y}{\dfrac{E}{2\alpha(1+\nu)} - \sigma_0} \right). \tag{7.3}$$

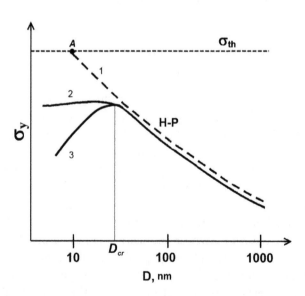

Fig. 7.1. Dependence of the yield stress σ_y on the average grain size D in a single-phase nanocrystalline material; 1 – the Hall–Petch relation (H–P); 2, 3 – experimental dependences in the area of violation of H–P at $D < D_{cr}$; σ_{th} is the theoretical strength.

Numerous experimental data [3] indicate that the decrease of the grain size in the nanoscale region below 50 nm results in a large deviation of the $\sigma_y(D)$ dependence from the H–P relation. Figure 7.1 shows schematically the experimentally observed dependences $\sigma_y(D)$ for polycrystalline nanocrystals: transfer to saturation (curve 2) and even a decrease of σ_y with a decrease of D (curve 3). Thus, the observed anomaly (deviation from the H–P relation) does not make it possible to obtain superhigh values of strength, close to σ_{th}, which are found in the solid.

The anomaly of the dependence $\sigma_y(D)$, like other special features of plastic deformation and fracture of the nanomaterials, is undoubtedly the consequence of the dramatic change of the structural mechanism of plastic deformation – activation in nanocrystals of the processes of low-temperature grain boundary sliding (LGBS) [4].

In a recent study [5] the principle of grain boundary engineering in the experimental verification of the proposed structural mechanism of plastic deformation using computer modelling methods was used to propose a new concept of the strengthening of nanocrystals based on the competition of dislocation slip and grain boundary sliding. The LGBS mechanism is more efficiently realised, as is well-known, in nanocrystals produced by controlled annealing of the amorphous state [6]. This is associated with the formation in late stages of crystallisation of thin grain boundary amorphous interlayers with higher thermal stability. Similar interlayers can be regarded as 'blurred' grain boundaries. Deformation of the nanocrystal in this case may take place by the LGBS mechanism by the formation of localised shear bands and 'blurred' grain boundaries. Examining the scheme in Fig. 7.1 it may be concluded that complicating the LGBS process may partially suppress the anomalous dependence $\sigma_y(D)$ and then move the nanocrystal to the level σ_{th} in accordance with the H–P relation (curve 1 in Fig. 7.1).

The most efficient method of influencing the LGBS process is, according to the authors of this book, the method of purposeful variation of the structure of the grain boundaries referred to in the literature as grain boundary engineering [7]. Actually, when examining the grain boundaries in the form of thin amorphous interlayers [6] one obtains:

$$\sigma_y = \frac{\sigma_0}{2}\left(1 + \frac{2L}{3\Delta}\right), \qquad (7.4)$$

where σ_y and σ_0 are respectively the yield stress and the stress for the nucleation of a shear band in the amorphous interlayer, L is the average grain size, Δ is the thickness of the grain boundary amorphous interlayer.

Equating the value L to the critical grain size D_{gr} at which an anomaly appears in the Hall–Petch relation (Fig. 7.1) gives:

$$D_{cr} = \frac{6\sigma_y}{\sigma_0} + 3\Delta. \qquad (7.5)$$

The grain boundary engineering method may be used in this case for increasing the value of σ_0 which should result, in accordance with (7.5), in a decrease of D_{gr} and, consequently, the possibility of obtaining higher strength of nanocrystals.

One of the methods of suppressing the LGBS process (increasing σ_0) is the formation of effective structural barriers in the path of propagation of deformation in the region of the grain boundaries. In [8] it was shown that the interaction of the propagating shear bands in the amorphous continuum with the nanocrystalline nanoparticles should result in efficient inhibition of plastic shear. By analogy, it may be assumed that in the presence the nanoparticles the processes of LGBs and 'blurred' boundaries will be inhibited. Thus, adding purposefully the nanoparticles in the region adjacent to the grain boundaries, in this stage of transition from dislocation slip to grain boundary sliding it is possible to increase the strength of the nanocrystals to the maximum possible values.

In order to confirm the validity of these considerations, in [5] it was attempted to obtain the discussed effect by computer modelling. The molecular dynamics method was used to study the change of the atomic structure of a Ti_2NiCu nanocrystal with precipitation of the second phase at the grain boundaries of the nanocrystal, under the effect of tensile stresses. Simulation was carried out using the Mie potential [9]. The Beeman–Schofield integration method, cyclic (toroidal) boundary conditions and the original cellular algorithm were used. The time was measured (with the accuracy to the multiplier 2π) in the periods of the Debye oscillations of the lattice. The unit of the mechanical stress in the internal system of units corresponded to the theoretical strength, and the normal elasticity modulus E in these units was equal to 30. The effect of the strengthening clusters of the second phase with the content of the light element (boron) of approximately 1 at.% was taken into account by defining the parameters of the interaction potential of the impurity atoms with each other and the impurity atoms with the main atoms.

Computer experiments in [5] were carried out in the isothermal conditions at a temperature of $T = 0.02$ which corresponds to 1/10 of the melting point of the main component. The behaviour of the system in the form of a polycrystal with a grain size of 10 nm and of the structural analogue was compared. The structural analogues contained in the most densely packed planes the layers of boron atoms for which the interaction energy with each other and with the main atoms was respectively 1.5 and 2 times greater than the interaction energy of the main atoms.

The computer simulation experiments showed [5] that the controlling factor in blocking dislocation slip or grain boundary sliding is the ratio of the equilibrium atomic spacings of the impurity in main atoms which in these computer experiments equalled 0.6 (experimental ratios for B with Ni, Ti, Cu are in the range from 0.56 to 0.66). Since the energy of interaction of the impurity atoms with the dislocations and the grain boundaries depends in a linear manner on the difference of the elasticity moduli and in the quadratic manner on the effective atomic diameters, the role of the atomic diameter of the impurity is one of the main factors.

The effect of tensile stresses of different magnitude in [5] resulted in plastic deformation (Fig. 7.2b, d). Figure 7.2a, b shows the

7.2a

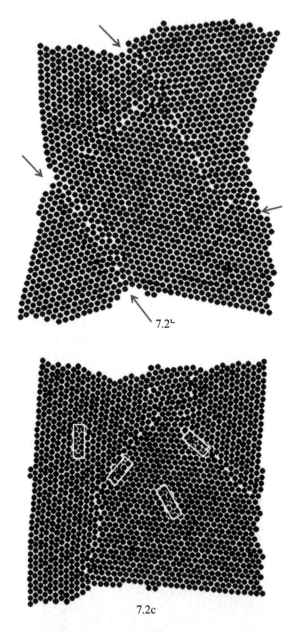

7.2ᴸ

7.2c

variation of the atomic structure of the alloy without strengthening clusters in the processes of plastic flow based on grain boundary sliding and the migration of the grain boundaries. The traces of sliding are indicated in Fig. 7.2b by the appearance of new steps and the shear of the nanograins in relation to each other.

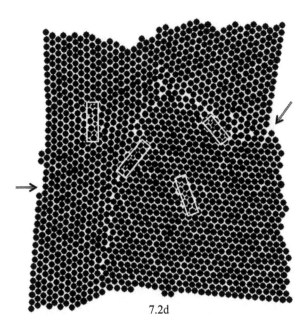

7.2d

Fig. 7.2. The atomic structure of the nanocrystal without strengthening clusters (a, b) and with strengthening clusters, indicated by rectangles (c, d) prior to (a, c) and after deformation by uniaxial tensile loading along the vertical axis (b, d); a, c − $\sigma = 0$; b, d − $\sigma = 1$, $t = 50$; the arrows indicate the stresses of sliding at the grain boundaries (b) and slip of the dislocations in the grain body (d).

Figure 7.2c, d shows the atomic structure prior to (c) and after (d) deformation, but it is shown for the system containing 1% of impurity clusters, with the half of the clusters distributed at the grain boundaries. It may be seen that the clusters efficiently block grain boundary sliding and grain boundary migration, and deformation developed mostly by the appearance and slip of the dislocations in the body of the nanograins. The rate of the process of grain boundary sliding decreases several times even in the presence of a single cluster. As indicated, this is manifested primarily in the blocking of the grain boundary sliding by the classes of the impurity (Fig. 7.2d). Thus, the results obtained in [5] shows that the clusters, containing the atoms with a small effective atomic diameter and with a high energy of binding with the main atoms effectively block grain boundary sliding and grain boundary migration thus stimulating dislocation sliding.

For the experimental verification of the concept of obtaining superhigh strength, described above, by means of the grain

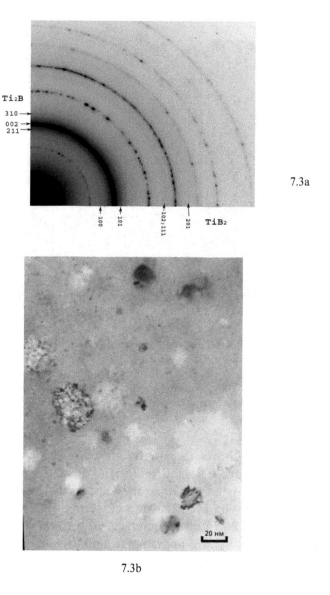

7.3a

7.3b

boundary engineering and heat treatment of the alloys $Ti_{50}Ni_{25}Cu_{25}$ and $Ti_{49}Ni_{24}Cu_{24}B_3$, produced by melt quenching, the authors of [10] produced structural states formed in late stages of controlled crystallisation and including, as shown by electron microscopic experiments (Fig. 7.3), a nanocrystalline matrix with the nanograins of the B2-phase (D = 30–40 nm) and with grain boundary nanoparticles of titanium boride with the size of 5 nm. The results show that the precipitation of the ultrafine phases of the grain

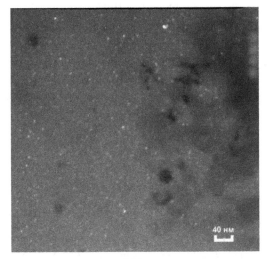

7.3c

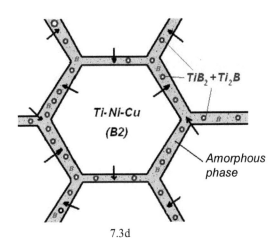

7.3d

Fig. 7.3. Structure of the nanocrystal of Ti$_2$NiCu in the superhigh strength state with the grain boundaries blocked by the nanoparticles of boron nitride with the size of approximately 5 nm; a–c – the transmission electron microscopy; a – electron microdiffraction pattern, b – bright field, c – dark field in the reflection of the the boride phase nanoparticles; d – diagram of the formed structural state.

boundaries for press the LGBS processes and displaces the value of D_{cr} to values lower than 40 nm. As a result of the experiments, it was possible to increase by 20% the normalised value of nanohardness [1] and displace it very close to the theoretical limit.

Thus, it may be seen that by theoretical considerations and computer simulation it has been possible to confirm the promising nature of the strengthening concept based on the composition of the processes of dislocation flow and grain boundary sliding in the nanocrystals. Undoubtedly, this would create suitable conditions for the purposeful control of the strength and superstrength of nanocrystals by grain boundary engineering.

References

1. Shtremel' M.A., The strength of the alloys, Part II, Deformation, Moscow Institute of Steel and Alloys, 1997.
2. Zhilyaev A.P., Pshenichnyuk A.A., Superplasticity and grain boundaries in ultrafine materials, Moscow, Fizmatlit, 2008.
3. Andrievsky R.A., Glezer A.M., *Usp. Fiz. Nauk*, 2009, No. 4, 337–358.
4. Kozlov E.V., et al., *Izv. RAN, Ser. Fizicheskaya*, 2009, V. 73, No. 9, 1295–1301.
5. Glezer A.M., et al., *Tech. Phys. Lett.*, 2016, V. 43, No. 1, 51–54.
6. Glezer A.M., Pozdnyakov V.A., *Nanostruct. Mater.,* 1995, V. 6, 767–770.
7. Glezer A.M., *Usp. Fiz. Nauk*, 2012, V. 182, No. 5, 559–566.
8. Glezer A.M., Shurygina N.A., et.al., Russian Metallurgy (Metally), 2013, No. 4, 235–244.
9. ASM Handbook, V. 22B. Metals Process Simulation (ed. by D.U.Furrer and S.L.Semiatin), Ohio, ASM International, 2010.
10. Glezer A.M., *J. Mater. Sci. Technol.*, 2015, V. 31, No. 1, 91–96.
11. Firstov S.A., et al., The new methodology formation processing and analysis of the automatic indentation of materials results, Kiev, LOGOS, 2009.

Conclusion

In this monograph, we have gradually examined a large number of dimensional effects accompanying the deformation behaviour of polycrystalline ensembles in metallic metals and alloys based on them. The size effect has a strong influence on the nature of the individual stages of the polycrystalline materials, on the defective (mostly dislocation) structure and the mechanical properties of nanocrystals, and also on the level of internal stresses. Important aspects of the nanostructured states must be taken into account when examining the pattern of high plastic strains. In the book, it is attempted for the first time to formulate a single comprehensive approach to this very interesting and important phenomenon and define this approach as a specific stage of the mechanical behaviour of any system, subjected to mechanical effects. On the example of titanium and its alloys it is shown that ion implantation is an efficient mechanism of nanostructural strengthening. Finally, the last chapter describes briefly an effective approach to the development of superhigh strength alloys with the mechanical properties approaching the theoretical strength. This approach, referred to in the literature as grain boundary design or grain boundary engineering, is based on the purposeful variation of the structure of the grain boundaries of the nanocrystals in accordance with the change of the susceptibility of the nanocrystals to low-temperature grain boundary microsliding.

We would like to stress that the information presented in this book relates mostly to the single-phase metallic nanomaterials and does not apply to the multiphase nanostates of the polycrystals. In order to fill this gap, we shall discuss briefly the promising two-phase amorphous–nanocrystalline alloys which are similar to the examined nanostructured materials.

The amorphous–nanocrystalline alloys is a new class of materials developed at the boundary between the 20th and 21st centuries as a result of the rapid development of new technologies and, in particular, nanotechnologies (various methods of producing amorphous and nanocrystalline powders and films, compacting, melt quenching, megaplastic deformation, implantation, laser, plasma and other methods of high energy fracture, etc) [1]. They were developed as a result of extensive research and investigations of advanced amorphous and nanocrystalline materials. The physical and mechanical properties of the two-phase amorphous–nanocrystalline materials are often superior to the properties of both nanocrystalline and amorphous materials resulting in a significant synergic effect. The special feature of the materials with the amorphous–nanocrystalline structure is the fact that the structural and phase components of such a two-phase system greatly differ in the nature of the atomic structure. In fact, the crystalline phase has an ordered atomic structure at large distances and is characterised by the presence of a translational symmetry. On the other hand, the atomic structure of the amorphous phase is disordered and has no translational symmetry of the long-range crystalline order and is characterised only by the distinctive topological and compositional short-range order in several coordination spheres. This 'unity of the oppositions'??? is not restricted to the above features and there are also a number of other large differences between the amorphous and crystalline states of solids.

The situation becomes even more unusual if the conventional crystal structural component is replaced by the nanocrystalline phase, characterised in particular by an additional set of unusual properties. In fact, the amorphous–nanocrystalline materials can be regarded equally as natural amorphous–nanocrystalline composites characterised by the important physical and mechanical properties for applications in practice [1].

Reference

1.   Glezer A.M., Shurygina N.A., The amorphous–nanocrystalline alloys. Moscow, Fiz-
      matlit, 2013.

Index